Microbial Physiology and Biochemistry Laboratory

Microbial Physiology and Biochemistry Laboratory

A Quantitative Approach

David White
George D. Hegeman
Indiana University

New York   Oxford
OXFORD UNIVERSITY PRESS
1998

Oxford University Press

Oxford New York
Athens Auckland Bangkok Bogota Bombay Buenos Aires
Calcutta Cape Town Dar es Salaam Delhi Florence Hong Kong
Istanbul Karachi Kuala Lumpur Madras Madrid Melbourne
Mexico City Nairobi Paris Singapore Taipei Tokyo Toronto Warsaw

and associated companies in
Berlin Ibadan

Published by Oxford University Press, Inc.
198 Madison Avenue, New York, New York 10016

Oxford is a registered trademark of Oxford University Press

Library of Congress Cataloging-in-Publication Data
White, David, 1936-
Microbial physiology and biochemistry laboratory: A quantitative approach / David White and
George D. Hegeman.
p.   cm.
Companion volume to: The Physiology and biochemistry of
prokaryotes.
ISBN 0-19-511313-6
1. Microbial metabolism—Laboratory manuals. 2. Bacteria—
Physiology—Laboratory manuals. I. White, David, 1936-
Physiology and biochemistry of prokaryotes. II. Hegeman, George.
III. Title.
QR88.W48 1995 Suppl.   97-8255
571 . 2'93—dc21   CIP

1 3 5 7 9 8 6 4 2

Printed in the United States of America
on acid-free paper

CONTENTS

INTRODUCTION

GOALS

The experiments in this laboratory manual illustrate major features of growth and metabolism discussed in *The Physiology and Biochemistry of Prokaryotes*.[1] Although the laboratory was designed to coincide with material discussed in the textbook, the sequence of experiments does not always coincide with the sequence of subjects presented in the textbook and the instructor is encouraged to schedule the experiments to coincide with the material discussed in lecture. Microorganisms studied include *Saccharomyces cerevisiae, Proteus vulgaris, Pseudomonas aeruginosa, Rhodospirillum centenum, Bacillus subitilis, Escherichia coli, and Photobacterium phosphoreum*. In addition to reinforcing what the students learn in lecture and from the textbook, the laboratory experiments described give students practical experience in several aspects of laboratory work, including:

- protein assays
- RNA assays
- enzyme assays
- oxygen uptake measurements
- column chromatography
- gel electrophoresis
- gas-liquid chromatography (GLC)

In addition, the students learn:

- how to process quantitative data and present them in tables and figures
- how to plan experimental procedures with regard to routine manipulations
- how to write laboratory reports (described in Appendix A).

We have assumed that each laboratory period is approximately 3 h, and have indicated the number of laboratory periods per experiment. Each experiment lists the minimum amounts of solutions and materials required per team of students. We have found that 2 students per team is optimal with regard to the students' learning experience. This can be readily done in laboratories of less than 15 students. In larger laboratories, teams of 3 may be necessary. All the organisms used can be obtained from the American Type Culture Collection and the ATCC numbers are noted where appropriate.

OVERVIEW OF THE EXPERIMENTS

There are 19 experiments that cover approximately 27 laboratory periods, 3 h each. We have found it useful to schedule separate discussion periods where the results of the experiments are discussed and new experiments are introduced. Six additional experiments, which are extensions of the laboratory experiments, are outlined in Appendix H as independent projects.

Experiment 1 is an introductory laboratory that is designed primarily for undergraduates who have had limited laboratory experience. It familiarizes the students with the pipettes, pipetting precision, and the colorimeters that they will be using during the course. They also examine the relationship between total cell density and turbidity (Beer-Lambert law). This can also be a discussion laboratory during which material in

Appendix A is introduced, the laboratory schedule and goals are discussed, and Experiment 2 is introduced.

Experiment 2 is a growth experiment where the students measure the different stages of growth of a population of bacteria using turbidity and viable cell counts. They also determine the relationship between turbidity and viable cell density.

Experiment 3 is a growth experiment with *Proteus vulgaris* during which the students perform a bioassay for nicotinic acid.

Experiment 4 is a growth experiment with *Escherichia coli,* in which the students learn the effects of environmental factors such as osmolarity, pH, and organic acids on growth yields.

Experiment 5 examines lactic acid production via the homofermentative and heterofermentative pathways. Students titrate the end products of fermentation.

Experiment 6 is an assay for RNA and protein in *E. coli* cells grown at two different growth rates. The students calculate the percent dry weight that is RNA and protein and the ratio of RNA to protein.

Experiment 7 is a study of diauxic growth of *E. coli* on glucose and lactose, in which the students measure growth rate, oxygen uptake, and ß-galactosidase.

Experiment 8 is an assay for amylase and protease secreted by *Bacillus subtilis.*

Experiment 9 is the concentration of amylase by ammonium sulfate precipitation as well as the separation of amylase from protease by affinity purification.

Experiment 10 employs ion-exchange chromatography to isolate amylase. The students construct their own columns and collect fractions either using a gradient maker or via step elution.

Experiment 11 is an assay for alkaline phosphatase in *E. coli.* Students make protoplasts and release the enzyme from the periplasm.

Experiment 12 teaches how to assay for threonine deaminase isolated from *E. coli* and how to determine the K_M and V_{max}.

Experiment 13 teaches the students how to do a coupled enzyme assay to measure the amount of fructose-1,6-bisphosphatase in extracts of *Saccharomyces cerevisiae.* Students also examine the effect of AMP, a negative allosteric effector, on the activity of the enzyme.

Experiment 14 is the partial purification of glucose-6-phosphate dehydrogenase from *S. cerevisiae.* This experiment teaches the students the techniques of gel filtration and gel electrophoresis.

Experiment 15 is the growth of *E. coli* anaerobically and aerobically in the presence and absence of nitrate and the assay for the induction of nitrate reductase.

Experiment 16 examines the cellular fatty acids of *E. coli.* The students extract the fatty acids, methylate them, and analyze them using gas liquid chromatography (GLC).

Experiment 17 is a study of chemotaxis in *Pseudomonas aeruginosa.*

Experiment 18 examines phototaxis in *Rhodospirillum centenum.*

Experiment 19 examines light production by the luminescent bacterium *Photobacterium phosphoreum.* The students perform experiments with metabolic inhibitors to examine the mechanism of light production.

Independent projects are described in Appendix H. The projects are related to the class experiments. Although guidelines are given as to how to proceed, students must read the cited references and devise their own protocols. One project is the determination of the amount of inorganic phosphate that limits the growth yields of *E. coli*. A second project is the investigation of factors that affect the relative amounts of saturated, unsaturated, and branched-chain fatty acids in *E. coli*. A third project is to measure the K_M and V_{max} of fructose-1,6-bisphosphatase, an enzyme that the class assays in Exp. 13. A fourth project is assaying for several glycolytic enzymes in yeast grown on glucose or ethanol. The students determine whether there are differences in enzyme activities dependent upon whether the growth substrate is glycolytic or gluconeogenic. A fifth project is a kinetic analysis of the regulation of fructose-1,6-bisphosphatase by AMP. This is a more detailed study of the inhibition by AMP observed in Expt. 13. Projects are also described using the luminescent bacterium *Photobacterium phosphoreum*, which is the subject of Expt. 19.

APPENDIXES

The appendixes include sections on:
- how to analyze experimental data
- how to do dilutions
- how to write laboratory reports
- quantitative problems and their solutions
- a list of laboratory supplies and equipment
- directions for making solutions
- independent projects

Appendixes A, F, and G should be studied by the students at the beginning of the course in order to prepare them for the calculations they will have to make during the laboratory exercises and the laboratory reports they will be expected to write. Mention of specific devices or products is made to provide examples the authors have reason to believe will serve the intended purpose. No endorsement of that device or product is intended.

ENDNOTES

1. White, D. 1995. *The Physiology and Biochemistry of Prokaryotes*. Oxford University Press, New York.

ACKNOWLEDGMENTS

We thank Eric J. White and Daisybrain Media Center for the illustrations, and Jim Bier, Carlos Ramirez-Icaza, Chamroeun Kong, and Darin Wolfe for aiding us in preparing this laboratory manual.

LABORATORY RULES AND SAFETY

The experiments you will be doing at times involve the use of toxic chemicals and bacteria that are opportunistic pathogens. In addition, some of the experiments require the use of high-voltage electrophoresis. Because of this, certain precautions must be taken. These are pointed out in the relevant sections of the experimental protocols, and will be summarized here.

CHEMICALS

In addition to strong acids and bases, you might use several potentially dangerous chemicals such as solvents, for example chloroform, toluene, and ether, acrylamide (a neurotoxin), and substances that are suspected to be carcinogens. You will be instructed not to mouth-pipette these chemicals, to use latex gloves, and, with respect to the solvents, to work in a fume hood. Needless to say, you may not eat, drink, or smoke in the laboratory.

MICROORGANISMS

The microorganisms with which you will be working are generally not pathogenic. It is best, however, to use careful microbiological techniques when handling the cultures. In particular, *Pseudomonas aeruginosa* is an opportunistic pathogen, and you should avoid exposure of cuts or burns to these organisms. Disposable plates of *P. aeruginosa* are to be discarded in the biohazard bags that will be made available, or if glass, placed lid up prior to autoclaving.

LABELING YOUR SAMPLES

It is important that you properly label all your samples that are placed on a shaker on in the incubator. The label should include your name, the course number, the date, and some mark of identification as to its contents. If you are labeling Petri plates, always label the bottom of the plate, that is, not the lid, so as to avoid confusion if the lids are interchanged.

BOOKS, NOTEBOOKS, COATS, ETC.

Do not place your personal belongings, such as books and clothing, on the laboratory bench. This interferes with the experimentation, and there might be chemicals on the bench that you wish to avoid.

Microbial Physiology and Biochemistry Laboratory

AN EXERCISE IN PIPETTING AND THE BEER-LAMBERT LAW

One Class Period

Goals

The goals of this introductory laboratory are to (1) introduce you to the use of various pipettes, automatic pipetters, and a colorimeter or spectrophotometer, (2) give you an idea of your pipetting precision, and (3) determine the relationship between cell density and absorbance.

INTRODUCTION

Throughout the semester you will be pipetting samples. The quantitative data that you obtain will be only as precise as your pipetting skills. For this reason, it is important to practice pipetting and to know the pipetting precision. An easy way to determine this is to make several replicate dilutions of a dye using a pipette and to measure the absorbance of the diluted sample. You will also determine the range over which the Beer-Lambert law, which describes the relationship between mass and absorbance, applies. (See Appendix E.)

MATERIALS

Supplies

- microliter pipettes that will deliver up to 100 microliters
- pipettes (1 and 10 ml)
- 16 x 150 mm test tubes

Equipment

- colorimeters
- phase microscopes

Solutions

Methylene blue. Stock solution is 1 mg/ml in water. Make 1 ml per team of students.

Growth medium for spectrophotometer blanks. Each team of students will need 5 ml.

Cells

A stationary-phase bacterial culture, such as *E. coli*. The instructor will determine the cell density by counting the cells in a counting chamber. Make 100 ml per team of students.

PROCEDURE

The goal is to pipette replicate samples (10) into test tubes and take readings with the colorimeter.

Determining the Variability in Pipetting

Using The Microliter Pipettes

Make solutions of methylene blue by diluting the stock methylene blue solution 1:400 into water with a microliter pipetter. If the diluent is 10 ml, what should be the volume of dye to make a 400-fold dilution? (Read the discussion of dilutions in Appendix A and do the dilution problems in Appendix F.) Do this 10 times and read the samples in the colorimeter. Make certain that you use only one colorimeter tube (because they vary) and mark the colorimeter tube so that it is placed into the machine in the same position each time. (Satisfy yourself that the reading changes when you turn the tube.) You can use the same pipette tip to make all of the replicate dilutions. However, carefully wipe the tip before you transfer the dye to the water to ensure that you do not accidently transfer dye solution that may be on the outside of the pipette tip. Examine the tip carefully to make certain that all the dye has been transferred to the 10 ml of water each time. It may be necessary to fill and empty the tip in the water in order to rinse out the inside.

Using the Regular Pipettes

1. Dilute the stock solution 1:40 into a test tube containing 10 ml of water.

2. Now do 10 identical tenfold dilutions by transferring 1 ml of dye into 9 ml of water for a final dilution of 400-fold. Use a 1ml pipette for thes dilutions.

3. Repeat the tenfold dilutions but use a 10 ml pipette. Fill the pipette with 5 ml of dye and ransfer1 ml to 9 ml of water.

Determining the Relationship Between Cell Density and Absorbance.

Students will dilute the culture in water using dilutions of their choice and determine the range over which the absorbance (turbidity) is linear with cell density. Dilute the samples 10%, 20%, 30%, 60%, 70%, and 80%. Measurements should be made of the undiluted and diluted samples at 660 nm (or 540 nm) and again at 440 nm using the same spectrophotometer or colorimeter. The blank should be growth medium appropriately diluted with water. Read the discussion of the Beer-Lambert law in Appendix E.

PREPARATION OF DATA

Record your results to indicate the variability. Suppose the readings were (% transmission): 37.4, 37.6, 37.3, 36.8, 36.9, 36.0. The mean (average) is 37.0.

One way to measure precision is to calculate the *variance* (S^2) which is the square of the standard deviation.

The formula is: $S^2 = \text{sum } (X_i - \bar{X})^2/N\text{-}1$, where N is the total number of measurements. In order to calculate the variance do the following:

1. Subtract the mean from each measurement and square the differences. For the example given, this would be 0.16, 0.36, 0.09, 0.04, 0.01, and 1.0.

2. Add the squares. Using the example given, this yields 1.66.

3. Divide the sum of the squares by N-1. The

example yields $1.66/5 = 0.332$.

The advantage to using the variance is that it is easy to calculate and treats all measurements equally; that is, it does not leave any out. The disadvantage is that it is a squared measure.

A second way to measure the variability is to use the standard deviation, (S), which is the square root of the variance. In this case, S is the square root of 0.332 or 0.576. One can expect that approximately 68% (2/3) of the measurements will be within one standard deviation of the mean and 95% will fall within twice the standard deviation.

A more complete discussion of error, including *probable error*, can be found in Appendix A.

Plot cell density versus absorbance at both wavelengths.

QUESTIONS

1. How precise are the microliter pipettes?

2. How precise are the standard pipettes?
3. Does it make a difference if one transfers 1 ml using a 10 ml pipette as opposed to pipetting 1 ml with a 1 ml pipette?

4. Compare the absorbance (or % transmission) of the dilute solutions prepared with the pipettes and the micropipette. Do they agree?

5. At what point did cell density become nonlinear with absorbance? Was this dependent upon which wavelength was used?

6. What is the numerical relationship in the linear range that you determined between cell density and absorbance units for each of the wavelengths? (See Table E.1 in Appendix E.)

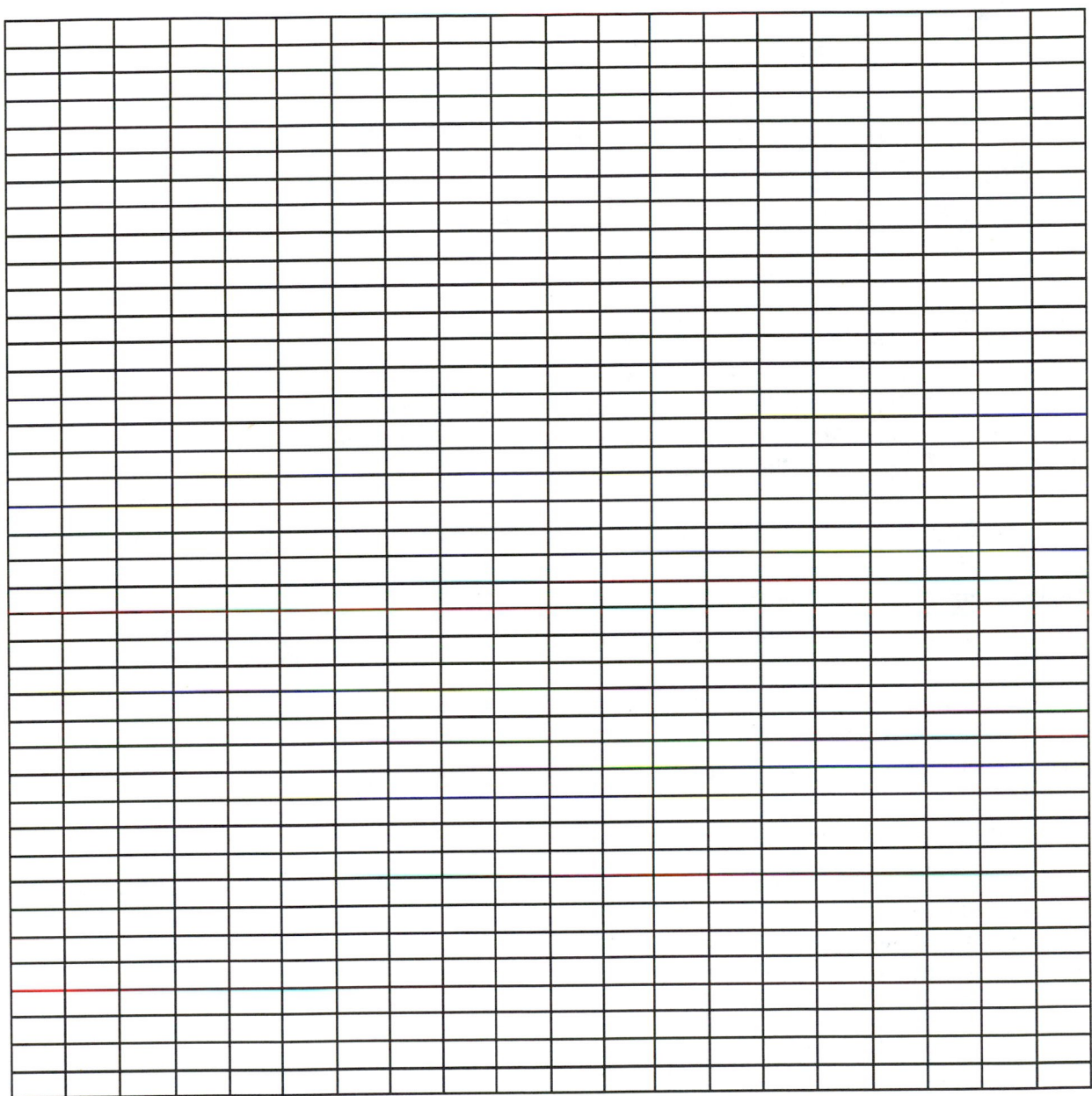

Experiment 2

BACTERIAL GROWTH CURVE

One Class Period

Goals

The goals of this experiment are to measure the different stages of growth of a bacterial culture using turbidity and viable count measurements, and to determine the number of viable cells/ml corresponding to one absorbancy unit.

INTRODUCTION

This experiment must be studied before class so that you are prepared to determine quickly the required dilutions for the viable count. Review Appendix A for how to do dilutions and do the dilution problems in Appendix F. You are responsible for calculating the dilutions to be made.

Growth is defined as an increase in mass and can be measured using any property that increases linearly with mass. Usually one uses absorbance (turbidity) or cell count (either total or viable). See the discussion of the relationship between mass and turbidity (the Beer-Lambert law) in Appendix E. Cell number is not always a valid measure of growth. As cells leave the lag phase, they may grow larger before they begin dividing, and as cells leave the exponential phase, they may continue to divide after growth has slowed or ceased, becoming smaller. This may be seen as a displacement to the right of the cell number curve relative to the mass curve when the data are plotted (Fig. 2.1).

MATERIALS

Supplies

- Nutrient Agar plates, 30 per team of students
- sterile 1.0 and 10 ml pipettes
- Eppendorf pipettes and sterile tips to deliver 100 μl for spread plates
- sterile 16 x 150 mm test tubes
- glass spreaders

Equipment

- colorimeters set at 540 nm for measuring growth, and colorimeter tubes

- a 37°C shaking water bath

Solutions

Sterile Nutrient Broth for dilutions. Each team of students will need approximately 300 ml.

Media for growth curve. Each student team should have 50 ml in a 250 ml Erlenmeyer flask. This assumes that 3 ml samples will be removed for absorbancy readings and that the same samples will be used for plating.

6

Cells

Use an overnight culture of *E. coli* grown in Nutrient Broth shaking at 37°C. For example, inoculate 50 ml in a 250 ml Erlenmeyer flask with 50 μl cf an overnight culture. The culture will be in stationary phase the next day. Let the culture sit at room temperature for about 1 hour before the laboratory period to ensure that a lag period takes place when growth is measured. Students will use these cells to inoculate media for the growth curves.

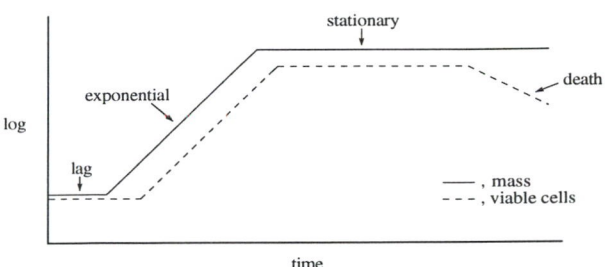

Fig. 2.1 A semilog plot of mass and viable cells during growth of a bacterial culture. (From White, D. 1995. *The Physiology and Biochemistry of Prokaryotes.* Oxford University Press, New York).

PROCEDURE

1. Measure the absorbance of the overnight culture at 540 nm. Zero in the machine with Nutrient Broth.

2. Transfer the culture to fresh medium, diluting the culture so that the absorbance is between 0.05 and 0.1.

3. Incubate in a shaking water bath at 37°C.

4. Sample the culture every 10 min during the lag period and measure the absorbance. The lag period might last 60 to 70 min. Do not forget the 0 min sample. Do not leave the culture on the laboratory bench after sampling. It must be returned to the shaker as quickly as possible.

5. Sample at 30, 60, 90, 120, and 180 min for both absorbancy measurements and plate counts. If you serially dilute the samples and plate 0.1 ml of the 10^{-4}, 10^{-5}, and 10^{-6} dilutions in duplicate, the cells plated should fall within a range that can be counted. Incubate the plates at 37°C for 48 h. Then the plates can be refrigerated until it is convenient to count the colonies.

PREPARATION OF DATA

Plot the log of the absorbance versus time as well as the log of the viable cells versus time. Calculate the length of the lag period and the doubling time. Calculate the number of viable cells per one absorbance unit at the wavelength used.

QUESTIONS

1. What have you learned about the relationship between absorbance and cell number from this experiment and Expt. 1? How does this vary with the wavelength?

NOTES

Experiment 2. Bacterial Growth

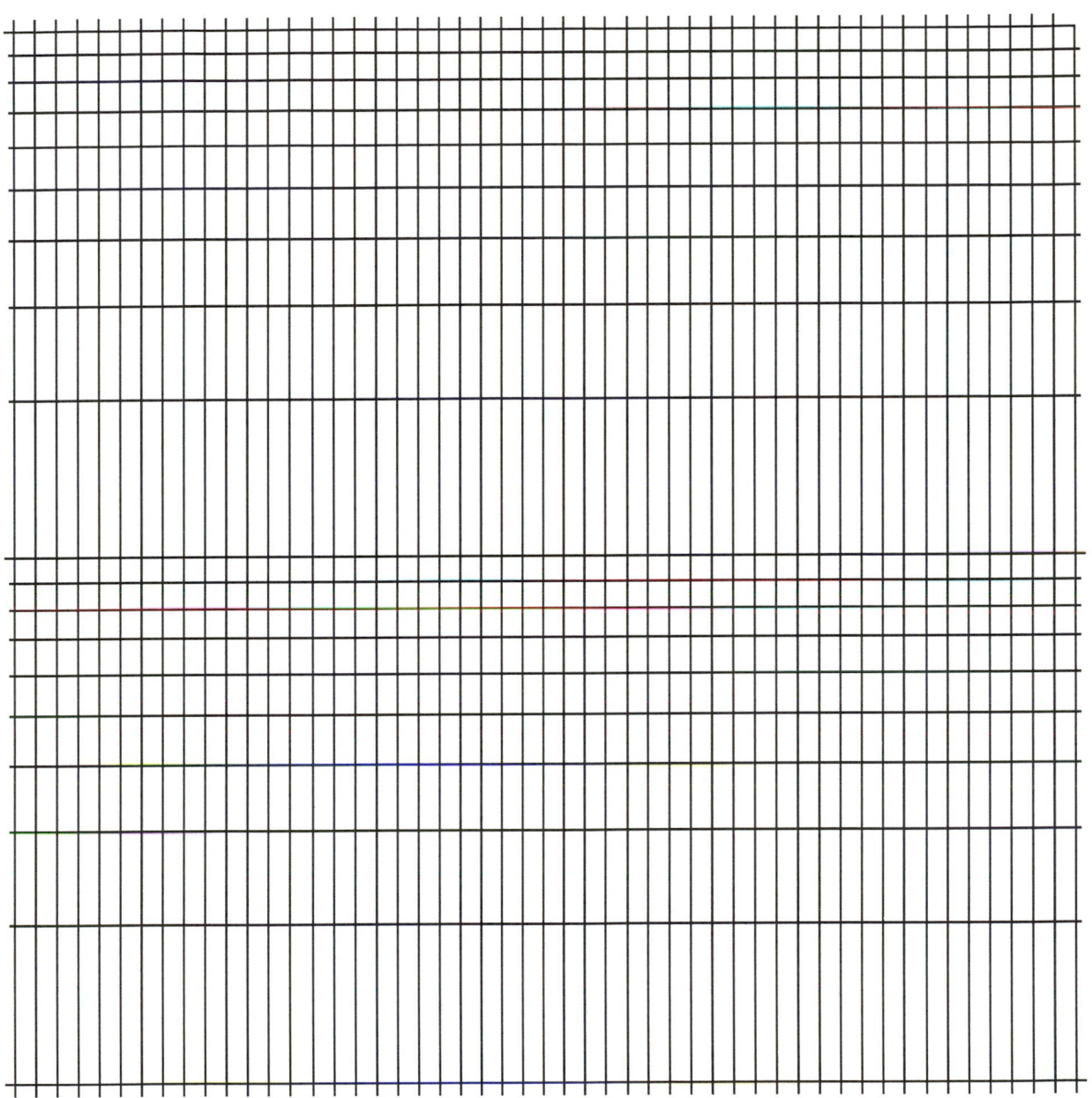

Experiment 3

BIOASSAY FOR NICOTINIC ACID

One Class Period

Goals

The goals of these growth experiments are to (1) give you experience in devising procedures for making direct and serial dilutions in order to arrive at a specified final concentration, (2) show you how growth yields can depend upon a limiting nutrient, in this case a vitamin, and (3) teach you how bioassays are performed.

INTRODUCTION

Nutrition is a key area of physiology. Through the study of the nutritional requirements of bacteria numerous growth factors have been identified. These have, in most cases, later been shown to be indispensable amino acids or components of cofactors that function with enzymes in key metabolic reactions. Since the metabolism of all cells has a common basis in evolution, these same microbial "growth factors" typically proved to function as well in higher organisms, including humans.

If an organism cannot synthesize a needed cellular component, it must be provided by the environment in which the cell is growing. This component then becomes a "growth factor" whose existence is revealed to the experimenter by the growth response (rate or amount of growth) to the added agent. This dependence of growth upon added growth factors is measurable in various ways and may be used as the basis for a "bioassay" for the growth factor in question.

Apart from their historical importance in the discovery of vitamins and amino acids, bioasassays are used today in the measurement of these substances in complex mixtures like foods and drugs. The advantages of bioassays are that they are highly specific, very sensitive, and economical to perform. All chemical forms of a vitamin that are effective precursors to a particular coenzyme are measured as one. The complexities of extraction and other aspects of sample preparation for chemical measurement are largely avoided.

The inhibition of growth as well as its augmentation may be measured by bioassay. A typical example is in the measurement of antibiotics or biocides like disinfectants.

Strains of bacteria or other microbes particularly useful in the bioassay of nearly every known growth factor are known and are listed in various sources, such as the catalog of the American Type Culture Collection. The online catalog plus search engine is http://www.atcc.org. (See *ASM News* **62**:605, 1996.) Commercial media useful in bioassays are available through various commercial sources (Difco, BBL, etc.).

Many strains of *Proteus vulgaris* require nicotinic acid, a precursor of NAD+ and NADP+, for growth. In this experiment, the application of *P. vulgaris* in the bioassay of this vitamin is shown by estimating the nicotinic acid content of a commercial preparation of yeast extract, a vitamin free hydrolysate of casein, and vitamin preparations. Yeast extract is often added to media for cultivation of bacteria as a source of growth factors such as the B vitamins. Vitamin-free casein hydrolysate is used as an amino acid supplement. A set of commercial vitamin preparations that each contain nicotinic acid are also available as "unknowns". The nicotinic acid content published on the label should be used as a guide for dilution of the preparation for addition to the medium; it should be appreciated that this published value is a legal minimum and that some variation from the published value may be expected due to incomplete mixing, degradation during storage, etc. Choose one unknown and calculate the dilution you need. A two-fourfold range (around 0.01-0.02 µg/ml of nicotinic acid) is a good idea, since you do not know exactly how much nicotinic acid is in the preparation. An instructor will help you filter sterilize the diluted sample for addition to the mineral medium.

MATERIALS

Supplies (per team of students)
- 22 sterile 16 x 150 mm test tubes
- 1 125 ml Erlenmeyer flask or beaker
- 1 100 ml graduated cylinder or volumetric flask
- 1 5 ml syringe
- 3 sterile 0.22 µm filters (Acrodisc)

- 1 spectrophotometer set at 660 nm (or Klett colorimeter, fitted with filter #66) and tubes for measuring cell density

Solutions (per team of students)
Culture medium. 160 ml SMB* containing 0.05% L-valine, 0.05% L-leucine, and 0.4% glucose, pH 7.0. Sterilize the glucose separately. It is convenient to prepare a 20% sterilized stock glucose solution and dilute it 1:50 into the growth medium. See Appendix C for formula for SMB*.

Nicotinic acid. Dissolve 2.0 mg in 200 ml water to obtain 10 µg/ml which the students will use for the standard curve. Filter sterilize or autoclave. Each team will need about 0.5 ml if assays are done in duplicate to construct a standard curve of nicotinic acid versus growth yield. The final concentrations of nicotinic acid will range from 0 to 0.2 µg/ml in a total volume of 5 ml of growth medium. There is no need to adjust the final volume to 5 ml, since the stock nicotinic acid is sufficiently concentrated that the volume of nicotinic acid is not a significant fraction of the total volume.

Yeast extract. A sterile 1 mg/ml solution of yeast extract. Students will add this to 5 ml to a final concentration of 0.005 and 0.01 mg/ml. Alternatively, each team receives 1.5 ml of a sterile solution of 0.1 mg/ml and must adjust the volume to 5 ml since the volume of the dilute yeast extract is a significant fraction of 5 ml.

Vitamin-free casein hydrolysate. A sterile 2% solution. Students will add this to 5 ml to yield a final concentration of 0.1%.

Vitamins. Use any vitamins known to contain niacin. If the vitamin is a tablet, crush the vitamin in water using an Erlenmeyer flask or a beaker as a container, and make an appropriate dilution based upon the information on the label. You will want the presumed niacin value to be within the standard curve of nicotinic acid. Filter sterilize the unknown as before using a syringe and Acrodisc.

Cells (per team of students)

Proteus vulgaris, Preceptrol culture(ATCC 13315, may be used. This strain has a requirement for L-valine and L-leucine in addition to nicotinic acid. The cells are grown 24-36 h to a high cell density in an Erlenmeyer flask at 30°C on a rotary shaker at approximately 200 rpm in SMB* (pH 7.0) plus 0.4% glucose, 0.05% each of L-valine and L-leucine, and 0.1 μg nicotinic acid per ml. Autoclave the glucose separately as a 20% solution and dilute it 50-fold into the sterile medium. The culture can be kept refrigerated for several days until needed. Each team will need 1.7 ml of a stock culture diluted 10^{-2}. It *is important to dilute the stock culture 10^{-2} before inoculating the unknowns in order to avoid transferring significant nicotinic acid from the growth medium to the experimental tubes.*

PROCEDURE

Preparing culture media

1. Using rigorous aseptic technique add 5 ml culture medium to each of 21 test tubes.

2. Add the following amounts of nicotinic acid (μg/ml, final concentrations) to each of 10 test tubes (use separate pipettes or tips for dispensing solutions): 2×10^{-1} (two tubes), 1×10^{-1}, 5×10^{-2}, 2.5×10^{-2}, 1.25×10^{-2}, 6.2×10^{-3} (two tubes), 3.1×10^{-3} (two tubes). The duplicate tubes will be read at a later time to make certain that growth had ceased at the time the first readings were made.

3. Add the following amounts of yeast extract (final concentrations) to each of two test tubes: 1×10^{-2} and 5×10^{-3} mg/ml Do these tubes in duplicate.

4. Add the vitamin-free casein hydrolysate to a final concentration of 0.1%. Do these tubes in duplicate.

5. Add vitamin unknowns to each of 4 test tubes Each team may test one unknown. Make two dilutions so that you are adding enough nicotinic acid to fall within the range of the standard curve. Do these tubes in duplicate.

6. One tube should have no additions.

Inoculation and incubation

Inoculate each tube with 100 μl of the diluted stock suspension of *P. vulgaris* and incubate at 30°C for 48-72 h shaking. Place the test tubes at a slant in a test tube rack and shake at approximately 200 rpm.

Measuring turbidity

After 48-72 h measure the turbidity at 660 nm. Measure the turbidity of the extra tubes at a later time (for example, in an additional 24-48 h) to ensure that growth has ceased. Use the same colorimeter tube for all measurements. Use the tube without nicotinic acid as the blank.

PREPARATION OF DATA

Draw a standard curve by plotting turbidity in the cultures against the initial amount (total in tube) of nicotinic acid. Construct a table similar to Table 3.1. See Appendix A for guidelines on constructing tables.

QUESTIONS

1. How much nicotinic acid is required to make 1 mg (dry weight) of cells? Assume that each cell weighs approximately 2.8×10^{-13} g and that there are about 1.5×10^7 cells/ml per Klett unit using the #66 filter.

2. How many molecules of nicotinic acid are there in each *Proteus* cell, assuming 1 dry cell weighs 2.8×10^{-13} g?

3. What compounds in the yeast extract or casitone other than nicotinic acid itself would you expect to be included in the nicotinic acid bioassay? Why?

4. What is the nicotinic acid content of your unknown (in units used on the label)?

5. Can you name other consequences of microbial growth (other than turbidity) that could be used to measure vitamins or amino acids?

Table 3.1 Bioassay of Nicotinic Acid

Tube	Yeast extract (μg)	Casein (μg)	Vitamin (tablet)	Turbidity (A_{660})	Nicotinic acid (μg)
1					
2					
3					
4					
5					
6					
7					
8					
9					
10					
11					
12					
13					
14					
15					
16					
17					
18					
19					
20					

NOTES

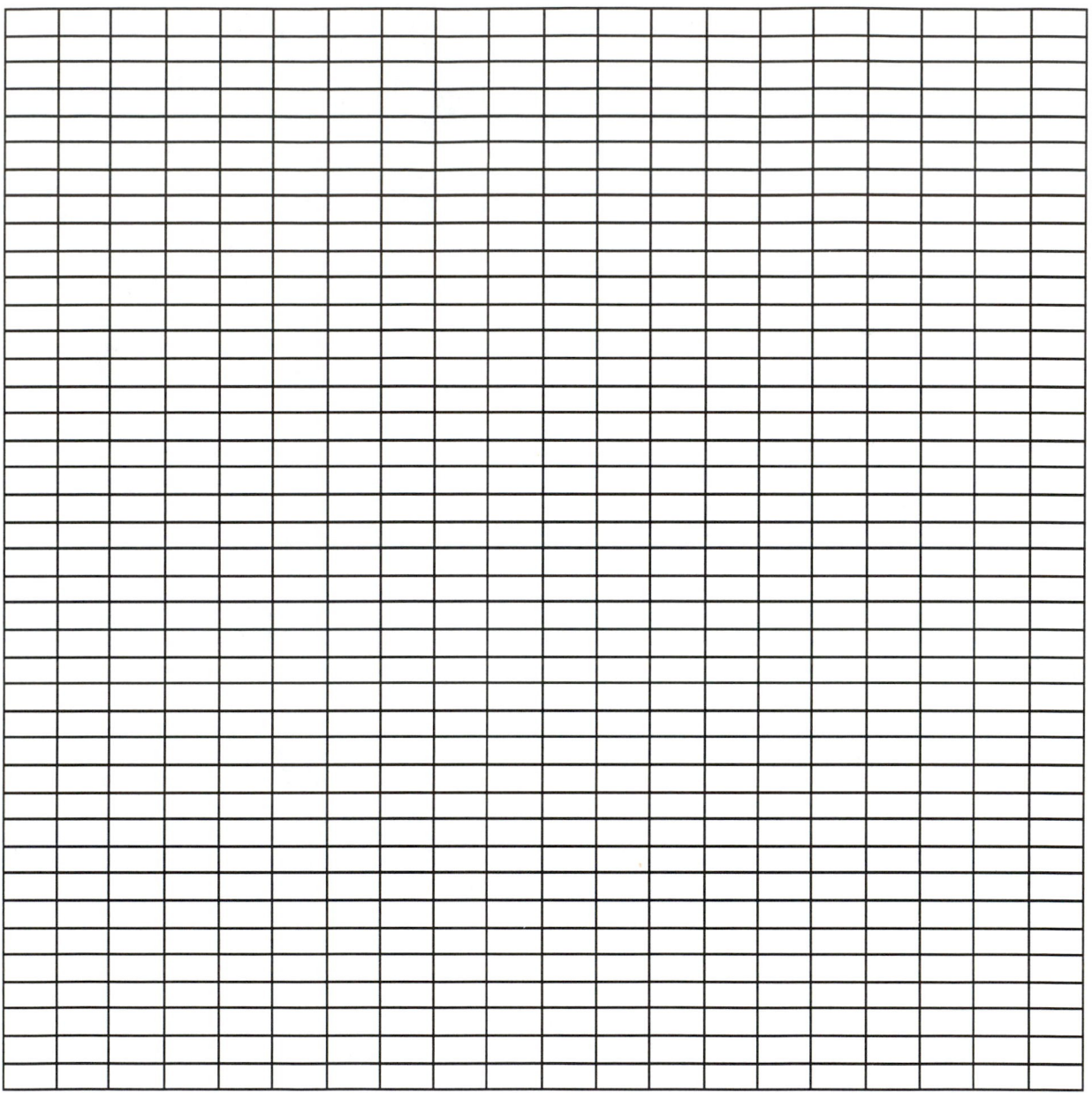

Experiment 3. Bioassay for Nicotinic Acid

Experiment 4

THE EFFECT OF ENVIRONMENT ON GROWTH

One Class Period

Goals
The goals of this experiment are to demonstrate how osmolarity, pH, organic acids, and sulfa drugs and their antagonists can affect the growth of bacteria. One of the lessons that you will learn is that organic acids can be inhibitory, but only at certain pH values.

INTRODUCTION

Although the construction of media for the cultivation of microbes is based on well-established nutritional and physiological considerations, unexpected effects of concentration and interactions among medium components can produce unpleasant surprises. Compounds that are good, readily used sources of carbon and energy at one concentration or at a given pH can become inhibitory at another concentration or pH. Compounds can be inert themselves yet can reverse the growth inhibition by a second compound or act with it to halt growth. In this experiment some of these effects are demonstrated using *Escherichia coli B*.

All the effects seen in this experiment have known chemical and physiological bases. In this deceptively simple experiment the student is invited to (1) use his/her knowledge of experimental design to describe the effects of medium variables and (2) do some reading to learn the mechanisms of the effects and to explain them.

MATERIALS

Supplies and Equipment
- Bunsen burners with striker flints
- colorimeters and tubes for measuring growth; turbidity can be measured at 660 nm.
- platform shaker at 30°C
- sterile pipettes
- sterile 16 x 150 mm test tubes

Media and Solutions

Medium A. sterile standard mineral base (SMB*) medium containing 0.05 M potassium phosphate buffer rather than 0.02 M potassium phosphate buffer and 0.1% NH_4Cl rather than 0.1% $(NH_4)_2SO_4$. Final pH 7.4. (See Appendix C.) Prepare at least 30 ml per team.

Medium B. Medium A + 40% (w/v) glucose. Prepare at least 6 ml per team. Filter sterilize rather than autoclave.

Medium C. Same as Medium A except that the final pH is 6.1. Prepare at least 10 ml per team.

Sterile water. 6 ml per team

Sterile sodium benzoate (1.2%), pH 6.5. 1 ml per team

Sterile sulfanilamide (50 mM). 1 ml per team

Sterile p-aminobenzoic acid, sodium salt (50 μM). 0.5 ml per team.

Cells

A fully grown culture of *E. coli B* inoculated from a fresh slant and grown shaking at 30^0C in 0.4% glucose mineral medium (SMB*, pH 7.4). Note: Media should be filter sterilized or autoclaved and glucose added separately. Each team will require 0.9 ml of culture.

PROCEDURE

Preparing Culture Media

Make up a series of 9 tubes according to the accompanying table.

Inoculation and Incubation

Inoculate each tube with one drop or 50 μl of culture and incubate shaking at 30^0C for two or more days. Make certain that the tubes are placed on a slant in the test tube rack.

Tube	Medium A	Medium B	Medium C	Benzoate	Sulfanilamide	PABA	Sterile H₂O	Growth (turbidity)
1	5.00 ml	-	-	-	-	-	1.0 ml	
2	5.00 ml	0.005 ml	-	-	-	-	1.0 ml	
3	4.95 ml	0.05 ml	-	-	-	-	1.0 ml	
4	-	5.00 ml	-	-	-	-	1.0 ml	
5	-	0.05 ml	4.95 ml	1 ml	-	-	-	
6	-	0.05 ml	4.95 ml	-	-	-	1.0 ml	
7	4.95 ml	0.05 ml	-	1 ml	-	-	-	
8	4.95 ml	0.05 ml	-	-	0.5 ml	-	0.5 ml	
9	4.95 ml	0.05 ml	-	-	0.5 ml	0.5 ml	-	

PREPARATION OF DATA

Record growth at 48 h using a colorimeter or photometer to measure turbidity. Check the tubes again at 72 h to ensure that growth has ceased. You may wish to verify the final pH values and examine the cultures as wet mounts to help you with your descriptions and explanations of the growth effects.

QUESTION

1. Describe and explain the effects of:
   a. Glucose concentration;
   b. pH;
   c. Benzoate at different pH values;
   d. Sulfanilamide and *para*-aminobenzoate.

NOTES

20   *Experiment 4. The Effect of Environment on Growth*

Experiment 5

LACTIC ACID PRODUCTION BY LACTIC ACID BACTERIA

One Class Period

Goals

This experiment is to (1) provide you experience in organizing and executing the quantitative measurement of a product formed attending bacterial growth, (2) show you how variation in a major metabolic activity within an important group of bacteria can be demonstrated, and (3) teach you quantitative skills in chemical measurement (titration) and calculation of physiologically important quantities deriving from that measurement.

INTRODUCTION

The production of particular fermentation products provides an accessible and reliable indicator of metabolic pathways used by bacteria. Quantitative or qualitative variations among products formed within an otherwise phenotypically homogeneous bacterial group can be of great diagnostic value.

The lactic acid bacteria ferment sugars and related compounds by several different pathways. The two simplest patterns, termed "homofermentative" and "heterofermentative", result in the formation, per mole of hexose fermented, of 2 moles of lactic acid or 1 mole of lactic acid, 1 mole of ethanol and 1 mole of CO_2, respectively (Fig. 5.1). Since CO_2 is volatile and is easily removed from acidic, aqueous solutions by boiling, and since ethanol is neutral, the difference in the amount of acid produced by cultures of these two fermentative classes from a given amount of hexose should be 2:1. This ex-

periment shows whether this is true and also measures in cultures of two lactic acid bacteria of the homofermentative and heterofermentative kinds the stoichiometry between hexose fermented and acid formed.

(The lactic acid bacteria are often used in bioassays of vitamins and amino acids. The extent of their growth in bioassay cultures is reflected in the amount of lactic acid they produce; hence, titration of acidity in such cultures provides a measure of the vitamin or amino acid.)

MATERIALS

Supplies (per team of students)

- 3 125 ml Erlenmeyer flasks
- 1 600 ml (or larger) beaker
- 1 Bunsen burner
- 1 ring stand and wire gauze support to heat water in the beaker
- 3 16 x 150 mm test tubes

A

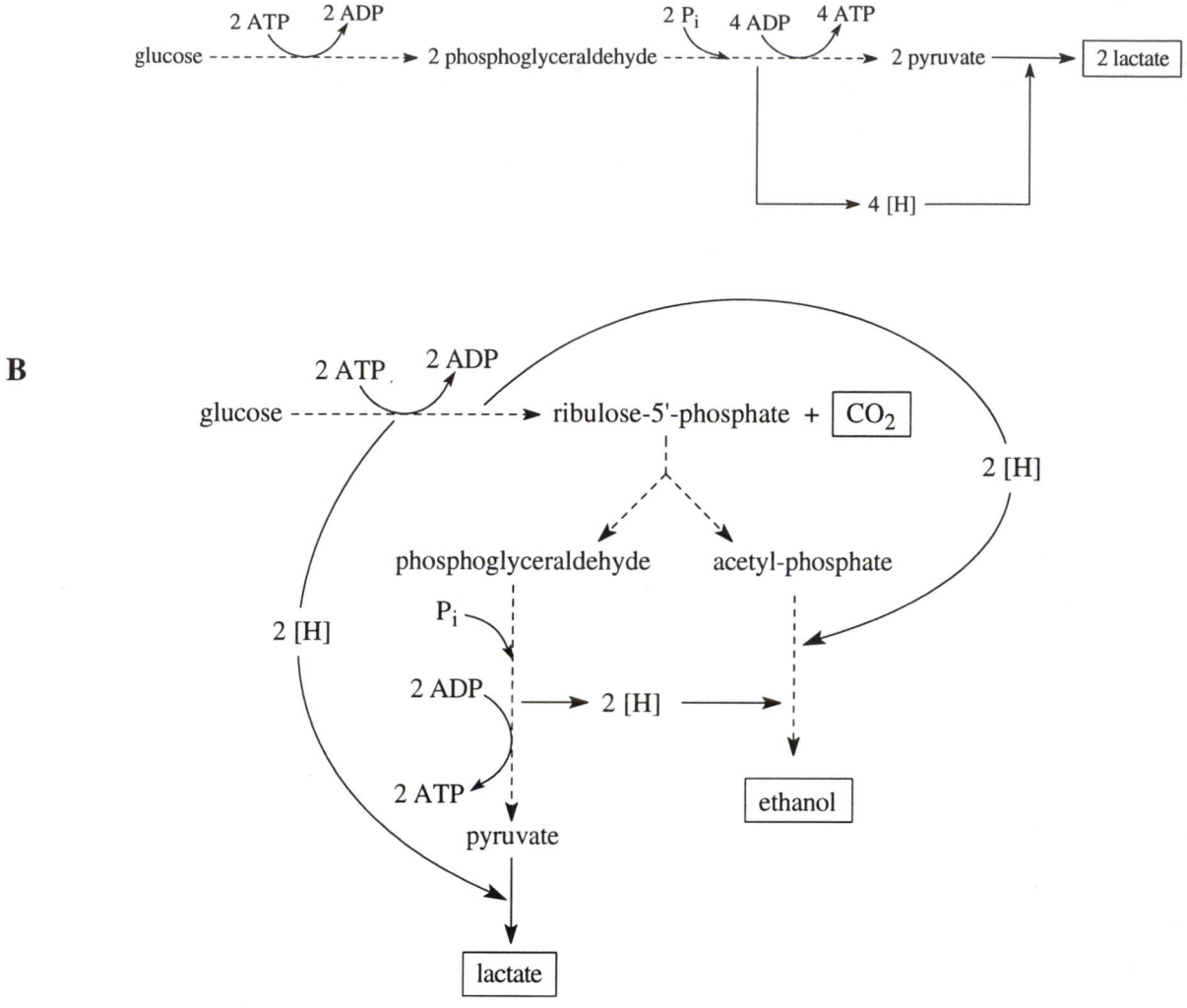

B

Fig. 5.1 Homofermentative and heterofermentative lactic acid pathways. A. The homofermentative pathway utilizes the Embden-Meyerhof-Parnas pathway and produces 2 moles of lactate per mole of glucose. B. The heterofermentative pathway utilizes the pentose phosphate pathway to produce ribulose-5'-phosphate, which is isomerized to xylulose-5'-phosphate, which is then cleaved to phosphoglyceraldehyde and acetyl phosphate. The phosphoglyceraldehyde is oxidized to pyruvate and then reduced to lactate, whereas the acetyl phosphate is reduced to ethanol via acetaldehyde. The products are carbon dioxide, ethanol, and lactate. Dotted arrows indicate that not all the reactions are shown.

- 25 ml measuring cylinder
- 25 ml burette or 1 10 ml serological pipette and pipette to aid to fit
- 1 test tube holder (clamp type)
- 200 ml sterile 1% tryptone, 0.5% glucose medium
- 2 60 ml ground glass stoppered bottles (sterile)
- boiling chips

Solutions (per team)

Culture Medium. 200 ml of 1% tryptone containing 0.5% glucose (unbuffered). This is conveniently prepared by mixing equal volumes (@ 100 ml) of 2% tryptone and 1% glucose (measure carefully) after autoclaving in a sterile 12 oz. screw-capped prescription bottle. This medium will later be poured into the 60 ml glass-stoppered bottles.

*1% Ethanolic Phenolphthalein.** Dissolve 1 gm of phenolphthalein in 100 ml of 95% ethanol and dispense in 30 ml dropper bottles.

*Benedict's Reagent.** Each team will use about 10 ml. See appendix C for formula.

0.1N NaOH (titrant). 30 ml of 0.1N NaOH made from standard (certified) 1 N NaOH.* Commercially available from e.g. J. T. Baker.

Cells (per team)

1 culture of homofermentative *Lactococcus lactis* (ATCC 11454)and 1 culture of heterofermentative *Leuconostoc mesenteroides* 8293). Prepare one fully-grown slant culture of each on ATCC medium no. 73 (YGC medium: 1% yeast extract, 2 % glucose , 2 % $CaCO_3$ and 1.5% agar) medium for each 6 students. Refrigerate until used. To prepare the inoculum for the students, inoculate

50 ml of the culture medium and grow for 2 days without shaking at 30⁰C. It is not necessary to grow the inoculum under anaerobic conditions.

* may be shared among several (e.g. 6) students conveniently.

PROCEDURE

Preparing cultures

1. Two weeks before the titration is to be performed each team should inoculate one of two filled, 60 ml sterile bottles with 50 µl of one of the two strains provided. It is more convenient to add the inoculum first and then to fill the bottle. Eliminate large air bubbles by carefully sliding the glass stopper in. Do not make the stopper tight in order to allow gas that may be produced during fermentation to bubble out. Incubate at 30°C on a tray to catch any spills. Save the remaining uninoculated medium in the refrigerator.

2. Before titration, test the 2 cultures for complete use of the glucose by transferring 0.3 ml of culture fluid from each separately to a 16 x 150 mm test tube, adding 3 ml of Benedict's reagent, then boiling each mixture briefly. Benedict's reagent measures reducing sugar (e.g., glucose) by reduction of cupric ion to an orange, colloidal precipitate of copper metal. The occurrence of such a precipitate in this test means that the culture should be incubated for an additional week and tested again at that time. Make up the volume of culture removed by adding fresh uninoculated medium.

   A negative test for reducing sugar is an unchanged blue or, at most, a pale green color after boiling. If the test for reducing sugar is negative, the titration may be done. If you have any doubts about the appearance of a positive Benedict's test, reconstruct a positive sample by use of 0.1 ml of uninoculated medium plus 0.2 ml of culture and repeat the test.

Titration

1. Using the measuring cylinder, carefully measure 25 ml of culture fluid from each of the two cultures into one of two 125 ml Erlenmeyer flasks. do not disturb settled cells. Measure 25 ml of uninoculated medium into the third. Rinse the cylinder with distilled water between uses. With 2-3 cm of tap water in the 600 ml beaker on the tripod, place each of the three flasks, in turn, into this water and boil each gently for 5 min to drive off CO_2. Use the test tube holder to remove the heated flasks and allow each to cool.

2. To each of the cooled flasks add 5-6 drops of 1% phenolphthalein. Holding a flask over a sheet of white paper, swirl it gently while adding the titrant (0.1N NaOH) from the burette (or pipette) dropwise until a stable, pink endpoint is reached. It is convenient for one student to swirl the flask while a second student add the NaOH drop by drop. Titrate the uninoculated medium first, and match the two cultures' endpoint colors to it. All cultures should give the same colors at the endpoint. Record the volume of 0.1N NaOH required to reach the endpoint for each of the two cultures and the uninoculated medium.

If you overshoot the endpoint in a titration, there should be enough culture as well as uninoculated medium left to repeat the tittration. Some students may wish to repeat

the titration of one or all cultures to see how repeatable the titration procedure is in their hands.

PREPARATION OF DATA

1. Subtract the amount of 0.1N NaOH required to reach the endpoint in the uninoculated medium from that required to reach the endpoint for each of the two cultures.

2. Assuming that the remaining titratable acidity is the lactic acid produced by fermentation of the glucose by the bacteria, calculate for each organism: (a) the number of moles of acid produced per mole of glucose fermented (m.w. of glucose = 180.16) and (b) the ratio between the number of moles of acid produced per mole of glucose by *Lactococcus* and the number produced by *Leuconostoc*.

QUESTIONS

1. The moles of lactic acid produced per mole of glucose fermented will probably be found to be a bit less than 2 for *Lactococcus* (or 1 for *Leuconostoc*). How do you account for the discrepancy if it occurred? Where, and in what chemical form, might the missing glucose be found?

2. What do you think are the major sources of error in this experiment? (Hint: How well could you judge the endpoint? What effect on the final color would the addition of a milliliter more [or less] of the titrant have made on this color? What effect would this additional milliliter have made on the calcu-

lated amount of acid per mole of glucose fermented?). How repeatable was the titration (for those who did duplicates)? How could the titration procedure be improved?

3. Why was it necessary to subtract the titrant used for the uninoculated medium from that used for the other cultures? What was being titrated in the uninoculated medium? Is it a fair procedure to do this subtraction?

4. Why was phenolphthalein ($pk_a = 8.5$) chosen as indicator? (This is not a simple question!)

NOTES

26

Experiment 5. Lactic Acid Production by Lactic Acid Bacteria

Experiment 6

ASSAY OF PROTEIN AND RNA IN WHOLE CELLS GROWN AT DIFFERENT GROWTH RATES

Two Class Periods

Goals

The goals of this experiment are to teach you how to assay protein and RNA in whole cells, and also to show you that the cellular protein and RNA content vary with the growth rate.

INTRODUCTION

RNA and Protein Vary With Growth Rate

When bacteria are grown at a fast growth rate, their cells are larger and contain more ribosomes per cell. The increase in ribosomes is due to the fact that, over a wide range of growth rates, ribosomes operate at an approximately constant efficiency, and their numbers in the cell limit the rate of growth. Thus faster-growing cells have more ribosomes. Because ribosomes are rich in RNA, the ribosome content is reflected in the cellular ratio of RNA to protein, which increases with growth rate. You will be given lyophilized *E. coli B* cells that were grown at different growth rates and asked to determine the percent of the dry weight that is RNA and protein and the ratio of RNA to protein. Fig. 6.1 summarizes published data on the macromolecular composition of *E. coli* grown at different growth rates.

Measurement of RNA

RNA will be extracted using hot trichloroacetic acid (TCA) and quantified by measuring the ab-

sorbance at 260 nm. (DNA also absorbs at 260 nm, but there will be very little DNA compared to RNA in the preparations.) The concentration in mg/ml of RNA = A_{260} / 23. An alternative method to measure RNA is the orcinol assay, which is described in Appendix D.

Measurement of Protein

The protein will be precipitated using TCA and measured using the Lowry assay. In the Lowry assay protein is reacted with the Folin reagent in an alkaline copper ion solution. A blue color develops due to the reaction of peptide bonds with the alkaline copper solution (the biuret reaction), and the reduction of phosphomolybdate-phosphotungstic acid in the Folin reagent by tyrosine in the protein. However, most of the color is due to the reduction of the phosphomolybdate-phosphotungstic acid by tyrosine. You will construct a standard curve using known amounts of bovine serum albumin. Plot the total amount of protein in the tubes on the abscissa and the absorbance on the ordinate. Then you can use the standardcurve to calculate the amount of protein in your sample after you have measured the ab-

27

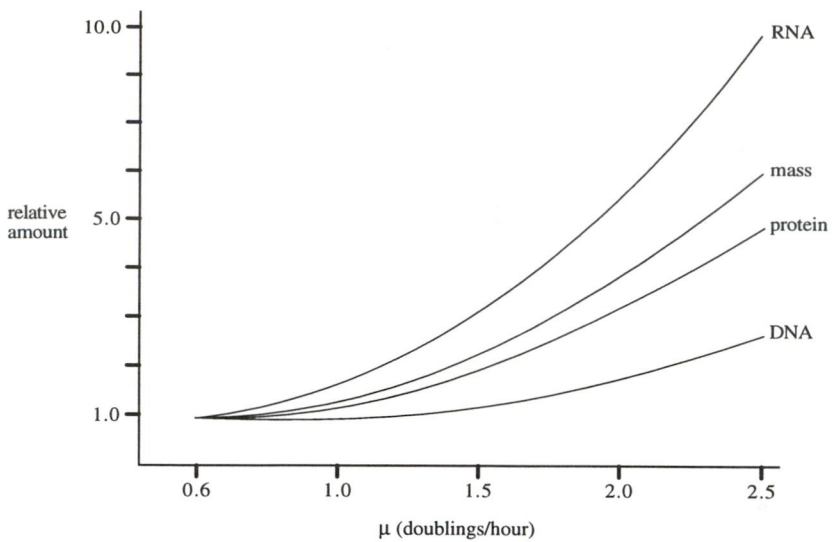

Fig. 6.1 Macromolecular composition of *E. coli* grown at different growth rates. The values are amounts per cell and have been normalized to values at a doubling time of 0.6 doublings/h. (From Neidhardt, F. C., Ingraham, J. L., and M. Schaechter. 1990. *Physiology of the Bacterial Cell.* Sinauer Associates, Inc. Sunderland, MA.)

sorbance of the sample.

MATERIALS

Supplies and Equipment

- water baths at 80 and at 50⁰C
- marbles to cover the tubes while in the water baths. Can use glass slides instead of marbles.
- spectrophotometer capable of reading in the uv range
- 1 ice bucket and ice per student team
- 1 15 ml corex centrifuge tube and rubber adapter per student team
- glass stirring rods. One per team of six students.

Solutions

1N NaOH. Will need 1 ml for each sample of cells analyzed.

10% (w/v) cold trichloroacetic acid (TCA) and 5% TCA. It is convenient to add water to a reagent bottle of TCA and make a 100% stock solution. The TCA solutions should be kept in the refrigerator. Will need 1 ml each of 10% and of 5% TCA for each sample extracted.

cold water. Will need 1 ml for each sample that is extracted.

Lowry assay. See Appendix D for directions on how to make the reagents. Each team will require approximately 40 ml of reagent C, 4 ml of the diluted Folin-phenol reagent, and 0.25 ml of the 1mg/ml BSA standard.

Culture Media

Minimal lactate. SMB* (pH 7.0), 1% sodium DL-lactate, 500 ml in a 2800 ml Fernbach flask (see Appendix C for SMB*).

Brain-heart infusion. Hydrated brain-heart infusion (e.g., Difco) medium, 500 ml in a 2800 ml Fernbach flask.

Cells

Lyophilized *E. coli*. Each team of 6 students should receive 8 mg in a 15 ml Corex tube. Grow the cells in either brain-heart infusion or minimal lactate. Grow 500 ml of cells in a 2800 ml Fernbach flask shaking at 200 rpm at 37⁰C. This will provide sufficient aeration for generation times of approximately 30 min (brain-heart infusion), and 60 min (minimal lactate) if the cell densities do not become too high. The cells should be harvested at a Klett (red filter) of around 100 - 110 while they are still growing at the maximum rate in exponential growth. See Appendix E for the turbidity readings using the green filter or a spectrophotometer at 660 or 540 nm. Growth begins to slow above a Klett #66 (red filter) reading of 100 for the lactate cells and around 120 for brain-heart infusion-grown cells. It is convenient to use an overnight culture diluted 1:100 as an inoculum. The initial Klett will then be about 4. Monitor the Klett readings. The cultures should reach a Klett of 100 within 2 to 8 h, depending upon the generation time and the lag period. Centrifuge the cells for 20 min at 8000 g. Resuspend the cells in cold distilled water and centrifuge at 12,000 g for 10 min. Freeze and lyophilize the cell pellets. Growth yields (dry weight) per 1000 Klett units are approximately 6 mg for brain-heart infusion-grown cells and 3 mg for lactate-grown cells (total Klett units = Klett reading x ml of culture) It should be noted that cells grown in brain-heart infusion grow as microscopic clumps, especially in late exponential phase. Clumping is so extensive that the clumps settle out of stationary suspensions.

SUMMARY OF PROCEDURE

The first step will be a cold trichloroacetic acid (TCA) extraction of the cells. This will be followed by centrifugation to yield a pellet and a supernatant fluid. (TCA precipitates the proteins and nucleic acids and leaves the soluble components such as the pool of free amino acids, sugars, and nucleotides in the supernate.) The pellet obtained by centrifugation will then be extracted with hot TCA followed by centrifugation. This will yield a second pellet and a second supernate. The nucleic acids and some polysaccharides are in the supernate of the hot TCA extract, and the proteins are in the pellet. You will perform a Lowry protein assay on the pellet and will determine the RNA content the supernate using the absorbance at 260 nm.

PROCEDURE

TCA Extractions

1. Weigh out 8 mg of washed, lyophilized cells and resuspend them in 1.0 ml of cold water in a 15 ml Corex centrifuge tube. You can use a glass rod to aid in resuspending the cells. Keep everything in ice. Some teams will assay cells grown in brain-heart infusion medium, and some will assay lactate grown cells.

2. Add an equal volume of cold 10% trichloroacetic acid (TCA) and let the sample sit on ice for 10 min.

3. Centrifuge the sample at 12,000 g for 20 min and pour off the supernate.

4. Use a glass stirring rod to suspend the pellet thoroughly in 1.0 ml of 5% TCA. Cover each tube with a glass marble and heat the suspension in an 80°C water bath for 30 minutes. This hydrolyzes the RNA and DNA but not the protein.

5. Cool and centrifuge the heated material for 20 min at 12,000 g. Very carefully remove the supernate with a Pasteur pipette. Record the volume of the supernate, and label the tube RNA. Add 1 ml 1N NaOH to the protein pellet and resuspend with a glass rod but do not shake. (Shaking might cause cell material to adhere to the walls of the test tube above the liquid.) Cap the tube with a marble, and place in a 50°C water bath while the RNA measurements are being done. Most of the TCA precipitate will go into solution. Measure the volume of the protein solution.

6. Calculate the expected concentration of RNA and protein in the samples (μg/μl)based upon the assumption that RNA can be10-20% of the cell dry weight and protein can be 30-60% of the dry weight.

RNA Assay Using A_{260} Readings

Dilute the RNA sample 1:100 with water and read at 260 nm using 0.05% TCA as the blank. Calculate the RNA content:

$$mg\ RNA/ml = A_{260} / 23$$

Multiply the mg RNA/ml by the total number of ml of TCA supernate to obtain the total amount of RNA in mg. Remember, you have diluted your sample 1:100 to take the readings.

Protein assay

1. Set up a series of 5 16 x 150 mm test tubes for the standard curve, one test tube for a reagent blank (everything but the protein), and 2 test tubes for each volume of sample to be assayed. The sample should be assayed in duplicate.

2. Add protein standards (a range of 20 to100 μg) and sample (a range of 20 to 400 μl). Use a volume of sample that delivers protein within the standard curve.

3. Add 4 ml of fresh reagent C to each of the test tubes and mix. (The blank has just reagent C.)

4. Incubate 15 to 30 min at room temperature.

5. Add 400 μl of Folin-phenol reagent:water (1:1) to all the tubes. Mix immediately after adding the reagent to each tube. You can do this by flicking the tube with your finger or by vortexing.

6. Incubate at least 45 min at room temperature. The color is stable.

7. Read absorbance at 660 nm.

Draw a standard curve for the protein assay. Plot the absorbance on the ordinate and the total μg of protein in each tube on the abscissa.

PREPARATION OF DATA

Make a table similar to Table 6.1.

Table 6.1 RNA and protein composition at different growth rates

Cells	Doubling time (min)	RNA (mg)	RNA (% dry wt)	Protein (mg)	Protein (% dry wt)	RNA/protein
brain-heart						
lactate						

QUESTIONS

1. What were the expected results of your experiment? How are these expected results rationalized in terms of cell physiology? What were the class results? Are your data in agreement with the class results and with the expected results? If not, offer some possible explanations. (Recall that the cells grown in brain-heart infusion grow as clumps. This may reflect extracellular materials e.g., polysaccharides, that might contribute to the dry weight.)

2. The color yield in the Lowry assay can vary from protein to protein when assaying purified individual proteins. What is the explanation for this?

REFERENCES

Johnson, J. L. 1994. Similarity analysis of rRNAs, pp. 683-700. In: *Methods for General and Molecular Bacteriology*. Gerhardt, P. (ed.). American Society for Microbiology, Washington, D.C.

Lowry, O. H., N. J. Rosebrough, A. L. Farr, and R. J. Randall. 1951. Protein measurement with the Folin phenol reagent. *J. Biol. Chem.* **193**:265-2

NOTES

32

Experiment 6. Assay of Protein and RNA in Whole Cells Grown at Different Growth Rates

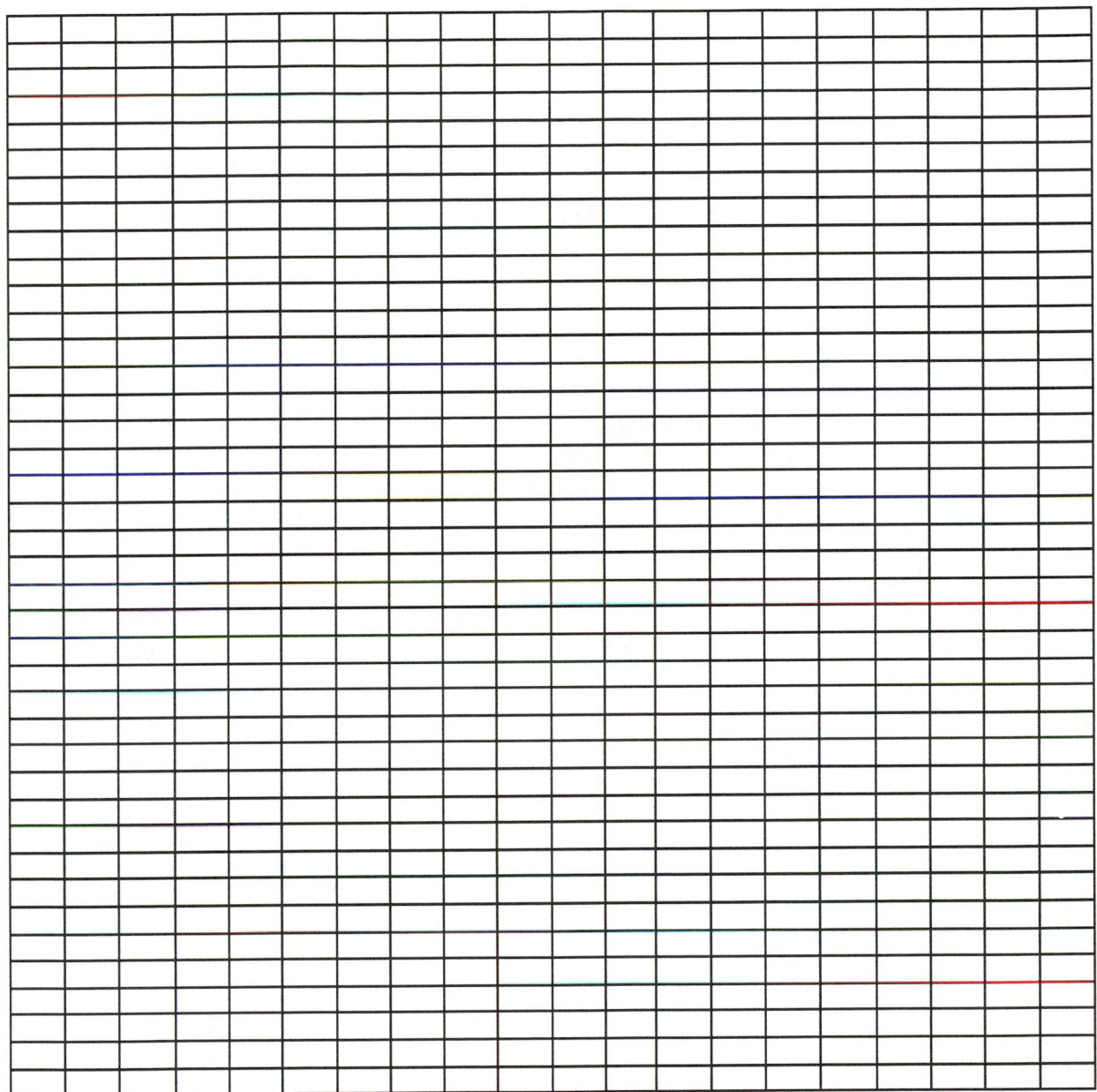

Experiment 7

ANALYSIS OF DIAUXIC GROWTH

Two Class Periods

Goals

The goals of this experiment are: (1) to demonstrate the phenomenon of diauxic growth and introduce you to glucose repression, (catabolite repression) (2) to show you how to measure the rate of oxygen uptake, (3) to teach you how to do a simple enzyme assay (i.e., the ß-galactosidase assay), and (4) to show you how to measure glucose.

INTRODUCTION

The phenomenon of diauxie (biphasic growth of a bacterial culture) was noted in the 1920s and 1930s. The selective use of one component of a mixture of sugars supplied as carbon and energy sources for growth at a time remained unexplained until the recognition of enzyme induction (derepression). The molecular genetic analysis of this phenomenon laid the groundwork for our current understanding of the regulation of enzyme synthesis in bacteria.

In its simplest form, diauxie is seen as two phases in a complex growth curve when two substrates are supplied. If glucose is one of the substrates in the mixture, it is typically used preferentially (first) by many (but certainly not all) bacteria—a phenomenon also termed the glucose effect. We now know that this actually reflects catabolite repression by glucose, in which the transcription of genes required for the utilization of the second carbon source is repressed when glucose is in the medium. Additionally, in *E. coli* glucose inhibits the uptake of other sugars, including lactose ("inducer exclusion"). Glucose repression in *E. coli* is due to the phosphotransferase system (PTS). During glucose uptake by the PTS system, the protein IIA_{gluc-P} donates its phosphoryl group ultimately to glucose to form glucose-6-phosphate. However, IIA_{gluc} stimulates the activity of adenylyl cyclase, the enzyme that makes c-AMP. Having donated the phosphoryl group to glucose, IIA_{gluc} no longer stimulates adenylyl cyclase; hence the cellular levels of c-AMP fall. This results in an inhibition of transcription of many genes, including those in the *lac* operon, because c-AMP and the c-AMP binding protein combine to stimulate the transcription of these genes. Additionally, the nonphosphorylated form of IIA_{gluc} inhibits the transporters of non-PTS sugars and thus their uptake. The net result, therefore, is diauxic growth. This laboratory exercise illustrates this phenomenon and its physiological basis.

Oxygen Uptake Experiments

You will be using a Clark oxygen electrode connected to a strip chart recorder (Fig. 7.1). The oxygen electrode consists of a Ag/AgCl reference electrode (a silver wire in a KCl solution) wrapped around a platinum electrode (Pt). The two electrodes are immersed in KCl and the tip is separated from the oxygen-containing solution by a membrane permeable to oxygen. An energy source maintains the Pt electrode about 0.8 volts more negative than the Ag/AgCl electrode, and a current flows, resulting in the electrolytic reduction of O_2 to H_2O at the Pt electrode (cathode) and the oxidation of silver at the silver anode. The current is proportional to the rate of oxygen reduction, which is proportional to the rate of diffusion of oxygen across the membrane. The rate of diffusion, in turn, is proportional to the concentration of oxygen in the solution outside the membrane. The current produced by oxygen reduction travels through a resistor and results in a voltage across the resistor according to Ohm's law: voltage = current x resistance ($V = IR$). The voltage is amplified and measured

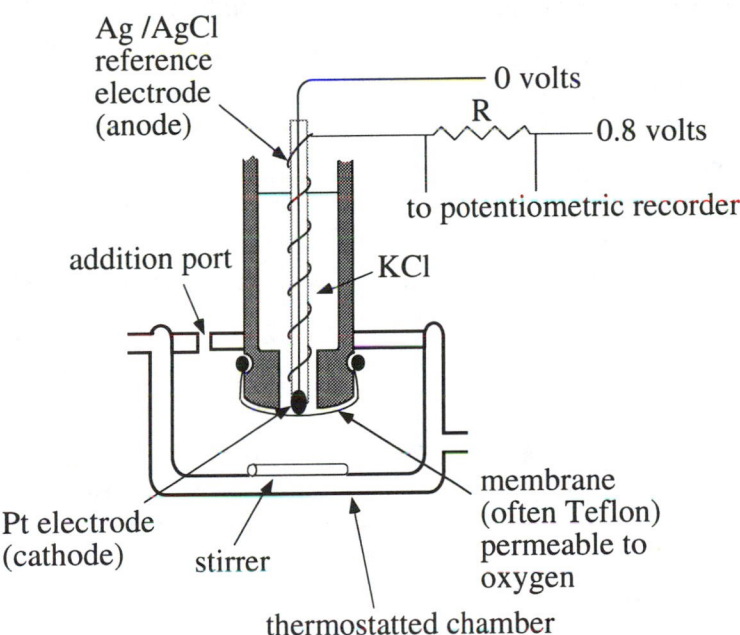

Fig. 7.1 An oxygen electrode. The platinum cathode is separated from the silver anode by an epoxy casing. The electrodes are in electrical contact via a KCl solution. A membrane permeable to oxygen separates the electrodes from the test solution or bacterial suspension. The membrane is attached via an "O" ring. A solid-state power supply maintains the cathode (Pt electrode) 0.8 volts more negative than the anode. Oxygen becomes reduced at -0.8 volts, and therefore, when it diffuses across the membrane to the cathode, it is reduced to water while the silver becomes oxidized at the anode. The current that flows is proportional to the rate of oxygen consumption (reduction) and is measured by a potentiometric recorder. The rate of oxygen consumption is proportional to the rate at which oxygen diffuses across the membrane, and this is proportional to the concentraton of oxygen in the suspension, that is,., outside the probe. Reading errors may occur unless the suspension is stirred because the oxygen can become depleted close to the membrane as the electrode consumes oxygen. (Adapted from Nicholls, D. G. and S. J. Ferguson. 1992. *Bioenergetics 2*. Academic Press, New York.)

with a potentiometric recorder. (See Fig. 7.1.) A bacterial suspension placed in a thermostatted, stirred cell in contact with the electrode is allowed to take up oxygen in the absence of added substrate. The rate of oxygen uptake during this time is the endogenous rate. A new rate is established and measured when a substrate is introduced. The endogenous rate is subtracted from this new rate to give a rate of O_2 uptake supported by the substrate.

Glucose assay

Glucose will be measured using a Glucostat kit. This is an enzymatic assay using a fungal glucose oxidase. The oxidation produces hydrogen peroxide. The hydrogen peroxide is measured by oxidizing a dye, dianisidine, catalyzed by the enzyme hydrogen peroxidase. When the dianisidine is oxidized by the hydrogen peroxide, a color is produced that is measured with a colorimeter or spectrophotometer. The reactions are as follows:

$$ \text{ß-D-glucose} + O_2 \longrightarrow \text{gluconic acid} + H_2O_2 $$

$$ H_2O_2 + \text{dianisidine (reduced)} \longrightarrow $$
$$ \text{dianisidine (oxidized-colored)} + 2H_2O $$

The two enzymes (glucose oxidase and peroxidase) are combined in a single reagent with the chromogen.

ß-Galactosidase assay

Cells will be assayed for ß-galactosidase, the enzyme that hydrolyzes lactose. The enzyme is assayed by use of the artificial chromogenic substrate, *o*-nitrophenyl-ß-D-galactopyranoside (ONPG), in which glucose is replaced by the aglycone, *o*-nitrophenol. The enzyme accepts this compound as a substrate in place of its natural substrate, lactose, and hydrolyzes the ONPG to galactose and *o*-nitrophenol. The *o*-nitrophenolate anion absorbs light strongly at 420 nm, and its hydrolytic release is readily measured since ONPG is colorless at this wavelength. ONPG does not enter cells readily enough to be useful in assaying ß-galactosidase in whole cells, but cells may be rendered permeable to it by use of a combination of an organic solvent (toluene) and a detergent (sodium dodecyl sulfate)

MATERIALS

Supplies and equipment

- Glucostat assay kit (Sigma Chemical Co.). Supplied with instructions for use.
- spectrophotometer at 420 nm and 450 nm
- Klett colorimeter with 660 nm filter
- Clark-type O_2 electrode connected to a strip chart recorder
- 13 x 100 mm test tubes for ß-galactosidase assay
- Eppendorf tubes for holding 1 ml samples
- 125 ml Erlenmeyer flasks for holding 5 ml samples
- a centrifuge for Eppendorf tubes
- ice buckets and ice to store samples
- chemical waste container
- syringes and needles for introducing substrate to oxygen electrode

Solutions

Glucose assay (per team). [Enough for 16 assays.]

Glucostat reagent. 80 ml.

SMB*, pH 7,4. 48 ml (see Appendix C).

glucose standard, 0.5 ml. 1 mg/ml glucose in SMB*.

ß-galactosidase assay (per team). [Enough for 12 assays.]

o-Nitrophenyl-ß-D-galactopyranoside (ONPG) 5 ml. Store refrigerated. 4 mg/ml in 0.1 M sodium phosphate buffer, pH 7. (See Appendix C for phosphate buffer.)

Toluene.

Sodium dodecyl sulfate (SDS), 0.1%.

Z buffer, 24 ml. 0.1 M sodium phosphate, pH 7, 10 mM KCl, 1 mM $MgSO_4$. Keep refrigerated. On day of use add 2-mercaptoethanol to 50 mM (350 µl per 100 ml).

Sodium carbonate (1 M), 12 ml.

O_2 uptake (per two assays)

SMB*, pH 7.4, 20 ml.

Lactose (glucose-free), 5% in water, 0.2 ml.

Glucose, 5% in water, 0.2 ml.

Cells

Escherichia coli B is grown overnight to stationary phase in 50 ml of SMB*, pH 7.4, glucose (0.04%), and lactose (0.2%). Autoclave the sugars separately and then add them to the sterile SMB*, or filter sterilize the medium. This culture is used to inoculate 100 ml of the same medium at a 1:1000 dilution in a 500 ml Klett-sidearm Erlenmeyer flask. The cells are grown at 37^0C shaking at about 180 rpm for 6 to 7 h until they reach a Klett #66 of 20. The generation time should be 50 to 60 min. At this point the culture is refrigerated overnight and used the next day for the diauxie growth experiment. If the students work in groups, each group should have a Klett flask of culture. It is important to note that the cells end growth on glucose and enter the transition period before growth on lactose when they reach a Klett reading of approximately 35. For this reason, it is important to refrigerate the cells when they grow to a Klett reading of 20.

PROCEDURE-DAY 1

Growth curve and oxygen uptake

Growing the culture. At the beginning of the laboratory period the turbidity of the culture in the 500 ml Klett-sidearm Erlenmeyer flask containing the culture at a Klett #66 of approximately 20 is read as soon as the flask is removed from the refrigerator and the flask is immediately placed on the shaker at 37^0C. This is the 0 min reading. There will be little or no lag period before growth resumes.

Taking Klett readings. Students should take Klett readings at time 0 and at 15 min intervals. At a Klett of approximately 30 to 40 the cells should switch from growth on glucose to growth on lactose. This will be reflected in a plateau of the growth rate and resumption of growth at a slightly slower rate. *You should be plotting the growth rate as you take the measurements so that you will know when this is taking place.* Students should be able to take at least 3 measurements (i.e., 0, 15, and 30 min) before the cells enter the transition period prior to growth on lactose.

Removal of samples for glucose analysis, O_2 uptake, and β-galactosidase assay. At the time each Klett reading is made, remove 1.0 ml of culture, place in a 1.5 ml Eppendorf tube, and keep on ice until it is convenient to centrifuge the sample. After centrifugation, remove the supernatant fluid and store it frozen. This will be used for glucose analysis in the next laboratory period. At two times, once during growth on glucose and once during growth on lactose, take 5 ml samples and place these in 15 ml Corex centrifuge tubes. Keep samples on ice until they can be centrifuged. These will be used for oxygen uptake experiments. At the same time that the 5 ml samples are taken, as well as other times of your choosing, take 1 ml samples and transfer to Eppendorf tubes and keep in ice. These will be stored frozen and used for the ß-galactosidase assay on day 2. For example, take samples during growth on glucose, during the transition period, during the early stage of growth on lactose, and during the later stage of growth on lactose.

Oxygen uptake. Centrifuge samples at 12,000 g for 10 min and wash with 5 ml cold SMB*. Resuspend in 5 ml SMB*, pH 7.4, in a small Erlenmeyer flask. Incubate the samples with shaking at 37°C for 15 min to starve the cells and reduce endogenous O_2 uptake. These cells will be used to measure glucose-stimulated and lactose-stimulated O_2 uptake.

Measure O_2 uptake using a Clark-type O_2 electrode connected to a strip chart recorder. When the suspension is in the electrode chamber and the air bubble is removed, record the endogenous rate of O_2 consumption (no substrate). Add lactose (0.1 ml of a 5% solution to 2.5 ml of cell suspension) next and record the rate. Subtract the endogenous rate. After recording the rate with lactose, add glucose (0.1 ml of a 5% solution to 2.6 ml of cell suspension) and record the rate. Subtract the endogenous rate. Make certain that you use separate syringes for each sugar. Subtract the lactose rate to estimate the glucose rate.

PROCEDURE-DAY 2

ß-Galactosidase analyses. The ONPG and Z buffer should be at room temperature. Set up a series of 13 x 100 mm test tubes to test each sample in triplicate, including a blank. You will be incubating each sample for 30 min. The rate of the reaction should be linear for this period as long as the absorbance does not get any higher than 1.0.

1. Add 2 ml of of Z buffer to each test tube.

2. Add about 2 Klett units (red filter) of cell suspension to each tube. [The number of Klett units is equal to the Klett of the cell suspension times the volume in ml delivered to the test tube (i.e., total Klett units = v x Klett). For example, if the Klett of the cell suspension is 100, then deliver 2/100 or 0.02 ml.]

3. Add 3 drops of toluene and 2 drops of 0.1% SDS to each tube and vortex vigorously for 3 sec. The vortexing is important to mix the

toluene and SDS thoroughly with the cells for maximum permeabilization of the cells.

4. Add 0.4 ml of ONPG to start the reaction to each tube except the blank.,

5. Add 1 ml of 1M sodium carbonate and 0.4 ml of ONPG to the blank.,

6. At 30 min, add 1 ml of 1M sodium carbonate to all the tubes except the blank. The increased pH inactivates the enzyme and also ensures total ionization of the phenolate anion for maximum color development. Mix.

7. Use the blank to zero the spcetrophotometer and take readings at 420 nm,

Enzyme activity can be expressed as a rate (absorbancy change per min) or specific enzyme rate (absorbancy change per min per Klett unit in the reaction tube.)

The supernate is assayed for glucose.

Glucose assay. Assay samples from day 1.

1. Add standards and sample to the test tubes. The standards should be 0, 20, 40, 60, 80, 100 μg of glucose. Recall that the supernate originally contained 0.04% glucose. Blank is the sample without glucose.

2. Add SMB* to a final volume of 0.5 ml.

3. Incubate the tubes for 30 min at 37^0C or 45 min at room temperature.

4. Measure the absorbance either with a Klett colorimeter with a blue (#42) filter or a spectrophotometer set at 450 nm.

Note: *Dianisidine is a member of a class of chemicals that may be expected to be carcinogenic (aromatic amines). Prudence requires that special precautions (pipettors, gloves) be taken in work done with it. Discard assay materials in the chemical waste jug provided before discarding glassware.*

5. Plot absorbance versus glucose concentration to prepare a standard curve. Determine the amount of glucose in the unknowns by using the standard curve.

PREPARATION OF DATA

Prepare a table with 6 columns (Table 7.1) showing (1) Time (2) Klett reading (3) Glucose (remaining in the culture medium) (4) Rate of glucose-dependent O_2 uptake per Klett unit (divide by Klett reading at the time the sample was taken) (5) Rate of lactose-dependent O_2 uptake per Klett unit, (6) Rate of ß-galactosidase per Klett unit in the assay tube. Make certain that the rates of O_2 uptake are corrected for endogeneous uptake.From this table prepare a graph which shows growth (semi-log plot), glucose levels, specific activity of ß-galactosidase (rate per Klett unit), and the lactose- and glucose-stimulated

Table 7.1 Glucose, oxygen uptake, ß-galactosidase, and turbidity readings

Time	Klett	Glucose (mg/ml)	Glucose-stimulated O_2 uptake (μl O_2/min/KU)	Lactose-stimulated O_2 uptake (μl O_2/min/KU)	ß-Galactosidase (A_{420}/min/KU)

oxygen uptake rates per Klett unit. The enzyme activities and oxygen uptake can be represented as bars.

QUESTIONS

1. Assuming that 1 Klett unit (red filter) = 1.5 x 10^7 cells/ml, how many cells (total) are produced form 1 mg of glucose in the growth medium used for this study? Assume that the dry weight per cell is 2.8 x 10^{13} g. What was the dry weight of cells produced in the growth medium per mg of glucose used? What per cent of the glucose was incorporated into cell material?

2. How does the onset of lactose-stimulated oxygen uptake and β-galactosidase activity correlate with glucose use?

3. Can you infer a cause-and-effect relationship of any kind from these data?

4. What conclusion(s) can be drawn from the oxygen consumption data alone?

5. What is the best way to plot the data relating β-galactosidase to growth? Why?

6. What are the sources of error in measurements and any other problems with this experiment, and what measures might one take to reduce or eliminate them?

REFERENCES

For a description of the oxygen electrode and measuring oxygen uptake, see Estabrook, R.W. 1967. Mitochondrial respiratory control and the polarographic measurement of ADP: O ratios. In: *Methods in Enzymology*, volume X, pp. 41-47. Estabrook, R.W. and M.E. Pullman, (eds.).Academic Press, New York.

For the β-galactosidase assay, see Miller, J.H. 1972. *Experiments in Molecular Genetics*, pp. 352-355, Cold Spring Harbor Laboratory, Cold Spring Harbor, N.Y.

NOTES

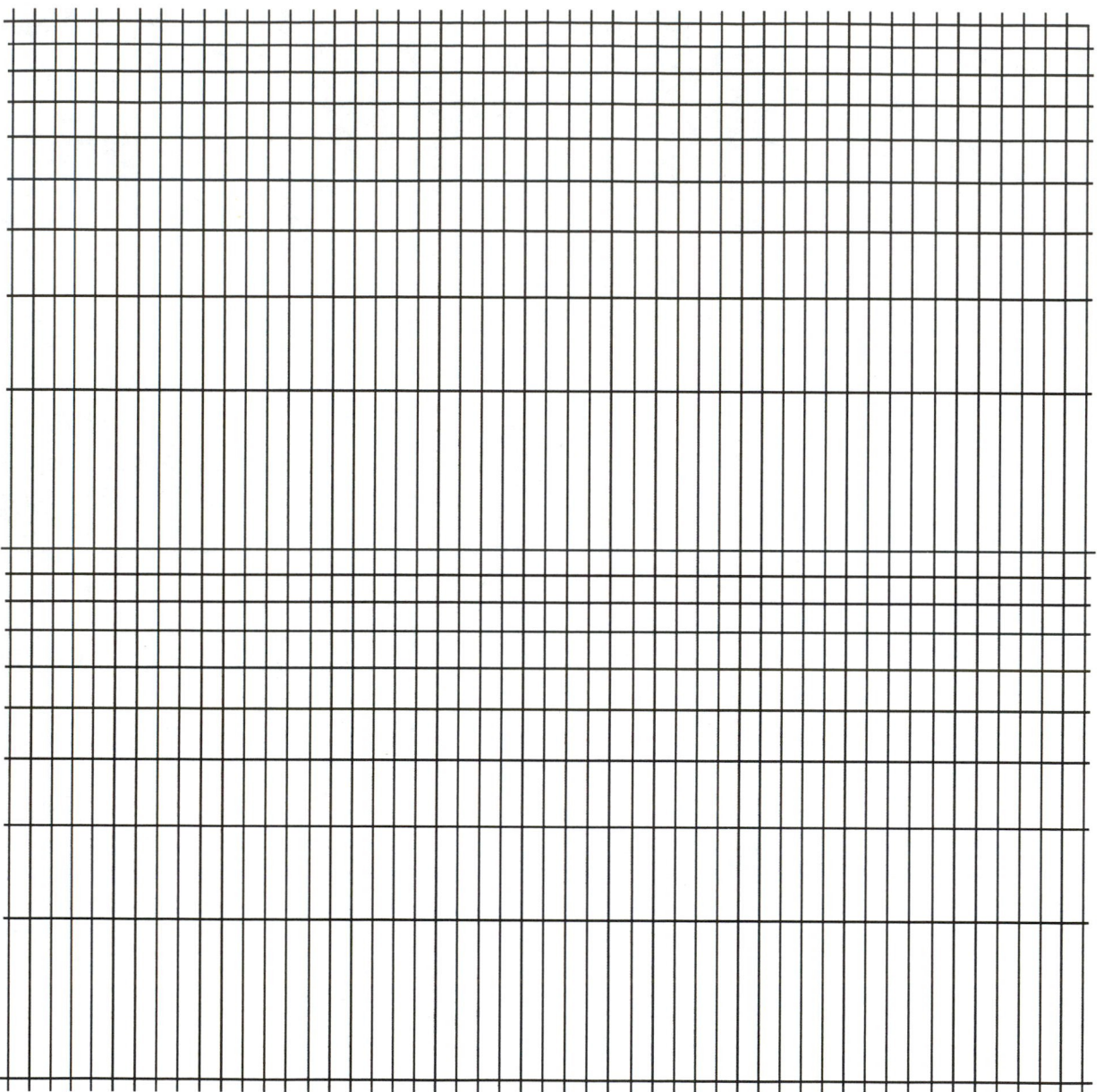

Experiment 7. Analysis of Diauxic Growth

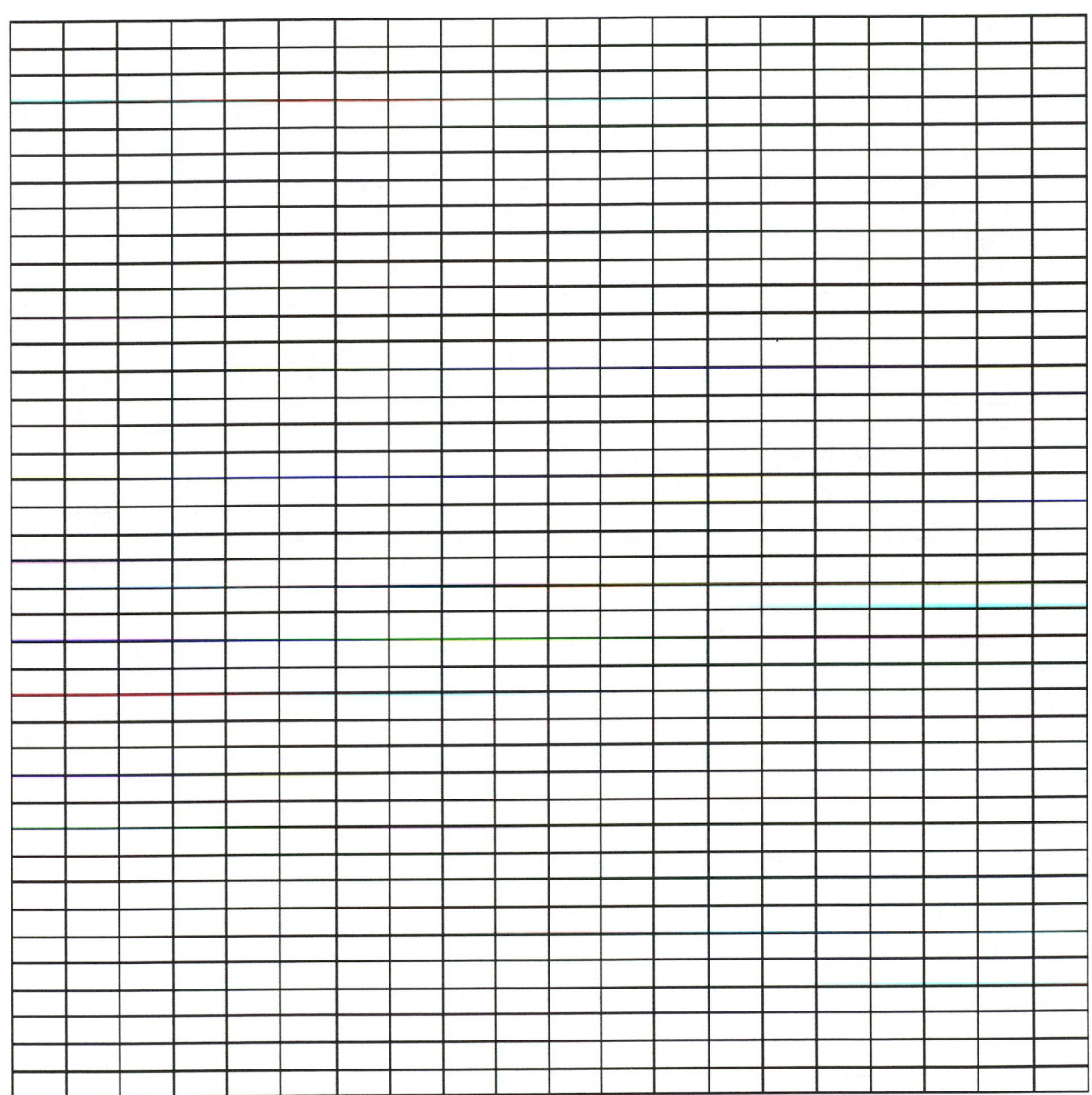

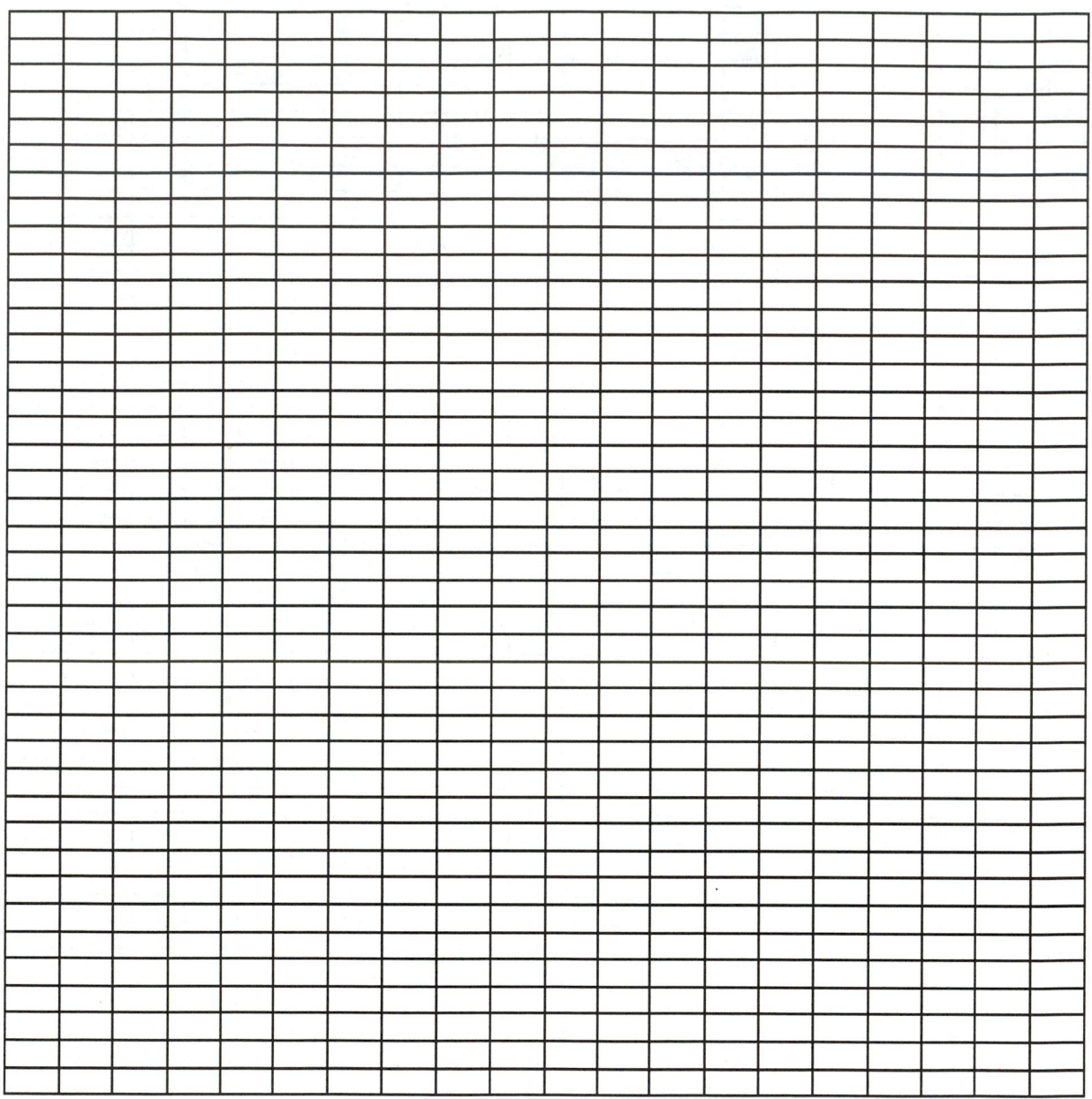

Experiment 7. Analysis of Diauxic Growth

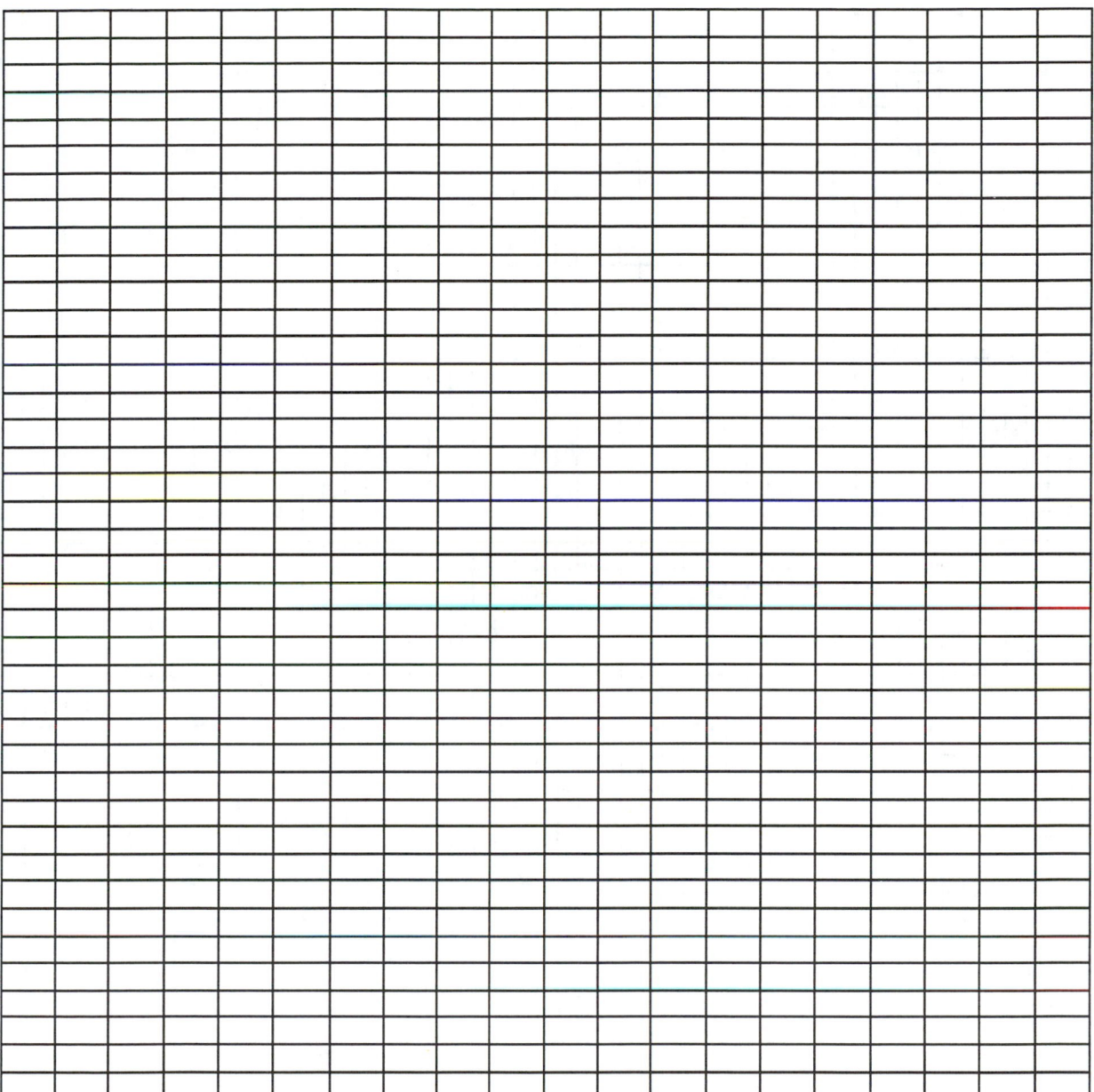

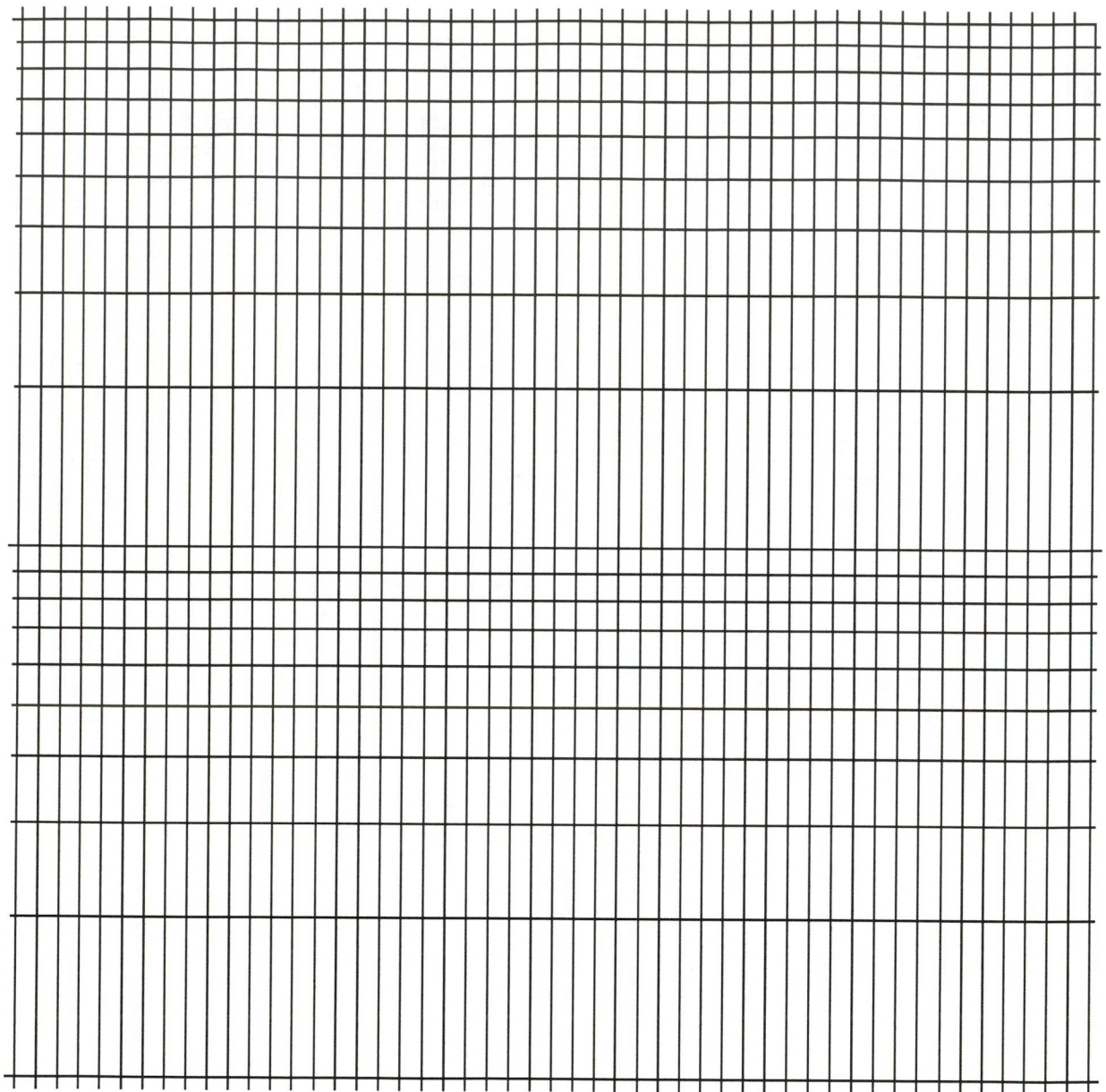

Experiment 7. Analysis of Diauxic Growth

Experiment 8

ASSAY OF AMYLASE AND PROTEASE SECRETED BY *BACILLUS SUBTILIS*

One Class Period

Goals

You will learn how to assay for amylase and protease and determine whether they are exoenzymes. This is a short experiment and can be combined with a discussion period or Experiment 9.

INTRODUCTION

Bacillus subtilis can utilize starch as a source of carbon and energy. In order to do this, the cells must secrete amylase to degrade the starch to smaller molecules that can enter the cells. In fact, certain strains of *B. subtilis*, such as the one you will be using, secrete so much amylase (and proteases) that the strains are used as a commercial source of the enzymes. The degradation of starch is monitored by the disappearance of material that stains with iodine. The assay detects α-amylase, which hydrolyzes α-1,4-glucoside linkages at random in starch, glycogen, and other polyglucosans. Intially, there is a rapid decrease in the molecular weight of the starch, resulting in a decrease in its iodine-staining properties. The final products are primarily small-molecular weight oligosaccharides. (In contrast, ß-amylases catalyze an exolytic attack and degrade starch by cleaving off maltose (a disaccharide) units from the ends of the starch chains.) The enzyme reaction is measured spectroscopically. The amylase is stable at room temperature and can be kept in the refrigerator for at least 48 hrs.

See Appendix D for an alternative assay for amylase. The protease assay you will use is a qualitative one. It is based upon the fact that ninhydrin reacts with the amino groups of alpha-amino acids and peptides to give a product with a blue color. The blue color develops as a result of chemical reactions during which an amino group is transferred to the ninhydrin which then reacts with a second ninhydrin molecule to give the blue product.

MATERIALS

Equipment

colorimeters set at 550 nm
water baths at 30°C and at 37°C
centrifuge and tubes for centrifuging the cells

Supplies (per team)

- 9 16 x 150 mm test tubes
- 4 13 x 100 mm test tubes

Solutions (per team)

KI (0.2% I_2 + 0.3% KI), 6 ml

2.0% Soluble starch in 0.1M potassium phosphate buffer, pH 6.5, 1.0 ml (Autoclave the starch for 15 min to put it into solution. If the starch solution is stored in the refrigerator, heat it e.g., in a steamer, for 5 to 10 min on day of use.)

0.1M Potassium phosphate buffer, pH 6.5, 30 ml. See Appendix C. (Will need more buffer for resuspending cells as described later.)

Bovine serum albumin (BSA), 1 mg/ml in water, 2 ml

Ninhydrin reagent, 3.5 gms in 100 ml of a 1:1 mixture of acetone and butanol. Each team will use about 2 ml.

Cells (per team)

Each team should receive approximately 10 ml of culture, which is a little more than they will use. *Bacillus subtilis* (ATCC 35854) is grown shaking at 30ºC into stationary phase in Nutrient Broth containing 0.2% soluble starch. Amylase production increases in stationary phase. The generation time should be approximately 75 min. It is convenient to inoculate each 50 ml in a 250 ml Erlenmyer flask with 50 µl of an overnight culture or a heavy inoculum from a plate or slant. Before harvesting remove 1 ml per team, and keep on ice. Label it *culture*. This will be assayed for amylase and protease. Centrifuge the remainder at 8000xg for 20 min. Transfer the supernate to another tube. Label this *supernate*. Each team should receive 1 ml of supernate for the amylaseand protease assays. Resuspend the cell pellet in 0.1M potassium phosphate buffer, pH 6.5, so that the cell density is about same as before centrifugation. Label this *cells*. Each team should receive 1ml of resuspended cells.

Students will assay the culture, the supernate, and cells resuspended in buffer.

PROCEDURE

Amylase assay

1. Pipette1 ml of the iodine reagent into each of 6 16 x 150 mm test tubes. Label the tubes 0'C(culture), 20'C, 0' S(supernate), 20' S, 0' cells, 20' cells.

2. Into each of 3 different 16 x 150 mm test tubes place 10 ml of phosphate buffer and 0.15 ml of the starch solution and mix. Label one tube *culture*, the second *supernate*, and the third *cells*.

3. Remove 2 ml from each of the 3 test and pipette into 1 ml of iodine reagent in the appropriately labeled 0 tube. Add 7 ml of water and mix. The color is stable and the absorbance can be read later.

4. Add 0.2 ml of enzyme (either *cuture*, *supernate*, or *cells*) to the remaining sample, mix, and place in 30ºC water bath.

5. At 20 min transfer 2 ml to 1 ml of iodine reagent, add 7 ml of water, mix, and read at 550 nm. Blank is water.

Protease assay

1. Add 0.4 ml of 1 mg/ml of BSA to each of 4, 13 x 100 mm test tubes. Label the tubes 0 (no enzyme), C (culture), cells, and S (supernate).

2. Add 0.2 ml of enzyme to each tube except the tube labelled 0, and incubate 30 min at 30°C.

3. Add 0.2 ml ninhydrin reagent, shake, place at 37°C for 10 min and observe color change.

PREPARATION OF DATA

Subtract the reading at 20 min from the reading at 0 min. and define an enzyme unit as a change of 0.1 A_{550} units per 20 min.

Make a table similar to Table 8.1 and record your data as units (U)/ml of enzyme for the amylase or +,- for the protease.

Table 8.1 Amylase

	Amylase[a] (U/ml)	Protease (+ or -)
Cells + supernate		
Supernate		
Cells		

[a]A unit (U) of amylase activity is 0.1 A_{550} units per 20 min

QUESTIONS

1. According to your data, where are the amylase and protease located?

2. Can you think of an explanation of why the protease does not destroy the amylase?

REFERENCES

Bernfeld, P. 1955. Amylases, α and β. Vol. I. pp. 149-158. In: *Methods in Enzymology.* Colowick, S. P. and N. O. Kaplan (eds.). Academic Press. New York.

NOTES

Experiment 8. Assay of Amylase and Protease Secreted by Bacillus subtilis

Experiment 9

CONCENTRATION OF AMYLASE FROM *BACILLUS SUBTILIS* BY AMMONIUM SULFATE PRECIPITATION AND SEPARATION FROM PROTEASE BY AFFINITY PURIFICATION

One Class Period

Goals

You will concentrate amylase by ammonium sulfate precipitation and also separate it from protease by adsorbing it to insoluble starch and eluting it with maltose.

INTRODUCTION

As shown in Exp. 8, *B. subtilis* secretes protease and amylase into the growth medium. The enzymes can be separated from each other by adsorbing the amylase to starch and subsequently eluting it with maltose. You will also concentrate the proteins by ammonium sulfate precipitation. The amylase will be assayed using the iodine assay described in Expt. 8. The protease assay is based upon ninhydrin. Ninhydrin undergoes complex chemical reactions with the amino groups from alpha-amino acids, primary amines, and peptides to form a purple product. The ninhydrin assay that you will be using is a qualitative rather than a quantitative assay.

Ammonium sulfate precipitation

Proteins can be concentrated or purified by ammonium sulfate purification. The process is called salting out of proteins. This is because salts at very high concentrations neutralize surface charges on the proteins and reduce the ef-

fective concentration of water. As a consequence the proteins interact with each other rather than with water and come out of solution. This can be used to purify proteins because the concentration of salt required to precipitate a particular protein reflects the number of charges and their distribution on the protein as well as other characteristics such as the number and distribution of hydrophobic amino acids that become exposed as the surface charges are neutralized, as well as the size and shape of the protein. At a sufficiently high concentration of ammonium sulfate, for example, 85% saturation at 0^0C, most proteins precipitate and therefore it is possible to concentrate bulk proteins from a crude extract using ammonium sulfate. You will do this to concentrate the amylase produced by *B. subtilis* (Expt. 8).

Separating amylase from protease

It is possible to separate the amylase from the protease by adding insoluble starch to the culture supernate to adsorb the amylase. The protease is not adsorbed. The amylase can be subsequently eluted from the starch with maltose (or soluble starch). Schwimmer and Balls were able to purify the amylase from barley using a similar procedure.[1] The technique of purifying a protein by adsorbing it to an insoluble ligand is called affinity purification and has been used to purify other proteins. The principle is that the protein is applied to an immobilized ligand to which the protein has an affinity. For example, the ligand may be an antibody to the enzyme or a molecule resembling the substrate, or a cofactor for an enzyme. If column chromatography is used, then the ligand is attached to a solid support (a resin or gel). See the article by Ostrove for a review of affinity chromatography.[2] The advantage to affinity chromatography is that the protein can be greatly purified in a single step from a crude extract. Since the amylase will attach to insoluble potato starch, which can be sedimented by centrifugation, the enzyme can be separated from the protease and purified without the use of a column.

MATERIALS

Equipment
- colorimeters set at 550 nm
- refrigerated centrifuge and tubes (50 ml) for centrifugation
- water bath at 30°C
- water bath at 37°C

Supplies (per team)
- 2 100 ml beaker
- 2 ice bucket and ice
- 15 16 x 150 mm test tubes
- 4 13 x 100 mm test tubes
- colorimeter tubes
- magnetic stirrers and stir bars to fit in the 100

ml beakers, 2

Solutions (per team)
KI (0.2% I2 + 0.3% KI), 15 ml

10% Maltose in 0.1 M potassium phosphate buffer, pH 6.5, 10 ml

2.0% Soluble starch in 0.1 M potassium phosphate buffer, pH 6.5, 1.0 ml (autoclave or steam the starch solution for 5 min to put it into solution)

0.1 M Potassium phosphate buffer, pH 6.5, 30 ml. See Appendix C.

Bovine serum albumin, 1 mg/ml in water, 2 ml

Ninhydrin reagent (3.5 gm ninhydrin in 100 ml of a 1:1 mixure of butanol and acetone). Each team will use about 1 ml.

Bovine serum albumin, around 5 mg

Insoluble potato starch, 5 g

Cells and enzyme (per team)
Each team should receive 50 ml of *Bacillus subtilis* (ATCC 35854). 10 µl of an overnight culture is inoculated into 50 ml of nutrient broth containing 0.2% soluble starch (autoclaved with the media) in a 250 ml Erlenmeyer flask *or* grown for 22 to 24 h using an inoculum from an agar plate. The culture is grown overnight shaking (approximately 180 RPM) at 30°C to stationary phase. The amount of enzyme increases significantly in stationary phase. The cells are removed by centrifugation at 8000 g in the cold for 20 min and the supernate (place in an ice bath) is

used for the enzyme assay and purification.

PROCEDURE

Precipitating the protein[3]
1. Transfer 25 ml of the culture supernate to a 100 ml beaker sitting in an ice bucket with a stir bar and slowly stir the solution for about 5 min to ensure that it is cool.

2. Add solid bovine serum albumin to a final concentration of 0.2 mg/ml. This is added as a carrier protein to ensure a significant precipitate.

2. Add 5 g of solid ammonium sulfate a little at a time. Wait until the amount you have added dissolves before adding the next increment. It should take around 10 min to add all the ammonium sulfate. This makes an 85% saturated solution at 0^0C.

3. Continue stirring in the ice bucket for 30 min or longer. [While it is stirring the amylase can be purified using the starch method.]

4. Centrifuge at 10,000 x g for 10 min. The supernates can be decanted.

5. Combine and resuspend the pellets with 5 ml of 0.1 M potassium phosphate buffer, pH 6.5, and keep on ice until ready to assay.

Purification of amylase using insoluble starch
1. Transfer 20 ml of the supernate to a beaker in an ice bath on a magnetic stirrer and start slowly stirring. The enzyme must be cold during the adsorption to starch so that the starch

is not degraded, which might result in the release of enzyme. Save1 ml of unused supernate for enzyme assay and transfer it to a tube in an ice bucket. Label it "C" for crude.

2. Add 4 g of insoluble potato starch to the remaining enzyme. Let it stir for 30 min so that starch does not settle to the bottom.

3. Centrifuge for 10 min in a refrigerated centrifuge at 10,000 x g to sediment the starch. (You can also centrifuge the ammonium sulfate precipitate at this time.)

4. Very carefully remove the clear supernate with a pipette. The starch pellet is not firm. It is all right if you leave just a little bit of supernate on the starch. Save the supernate and label it "U" for unadsorbed. You will assay it for enzyme that was not adsorbed.

5. Suspend the starch with 10 ml of 10% amylase and let it stir in a beaker at room temperature for 30 min. The enzyme binds to maltose and is released from the insoluble starch.

6. Centrifuge the preparation as before and save the supernate. Label it "P" for purified. This should contain the eluted purified enzyme.

Amylase assay
This assay will be done with the crude enzyme, with the unadsorbed enzyme, with the purified enzyme, and with the ammonium sulfate precipitated enzyme. The 4 preparations are assayed at the same time after the purification. Assay 50 µl of enzyme.

1. Pipette 1 ml of the iodine reagent into each of

8, 16 x 150 mm test tubes. Label 2 of the tubes C0 and C20 min. Label 2 of the tubes U0 and U20. Label 2 of the tubes P0 and P20, and 2 of the tubes N0 and N20. These tubes will used to assay the crude enzyme (C), the unadsorbed enzyme (U), the purified enzyme (P), and the ammonium sulfate precipitate (N).

2. Pipette 10 ml of the 0.1M phosphate buffer and 150 μl of the 2% soluble starch solution into each of 4, 16 x 150 mm test tubes. Label one tube C (crude enzyme), the second U (unadsorbed enzyme), the third P (purified enzyme), and the fourth N (ammonium sulfate).

3. Remove 2 ml of the buffer and starch solution and pipette it into the tube of iodine labeled 0 min. Mix. Add 7 ml of water and mix again.

4. Place the tubes labeled containing enzyme into a 30°C water bath and at 20 min transfer 2 ml into the appropriate tubes with the iodine reagent. Mix, add 7 ml water, and mix again.

5. Read the samples at 550 nm using water as a blank. The color is stable.

6. Subtract the absorbance at 20 min from the absorbance at 0 min. Define an enzyme unit as 0.01 absorbancy change per 30 min.

Protease assay

1. Place 0.4 ml of the 1 mg/ml bovine serum albumin in each of 4 13 x 100 mm test tubes. Label these tubes 0 (no enzyme), C (crude),U (unadsorbed), and P (purified).

2. Add 0.2 ml of enzyme to all of the tubes except the tube labelled 0.

3. Incubate 30 min at 30°C.

4. Add 0.2 ml of the ninhydrin reagent to all of the tubes and incubate at 37°C for 10 min. Solutions containing enzyme turn dark blue.

PREPARATION OF DATA

Calculate the % of amylase in the crude that was adsorbed to the starch, the % of amylase that was eluted from the starch, and the % of amylase precipitated by ammonium sulfate. For example, suppose 50 μl of the crude enzyme catalyzed an absorbance change of 0.76 units in 20 min. Then the number of units of enzyme per ml is 0.76/0.05 or 15.2 units per ml. If you started with 20 ml of enzyme, then you started with 304 units.

QUESTIONS

1. What % of the amylase adsorbed to the starch? What percent was eluted?

2. What % of the amylase was precipitated by ammonium sulfate?

3. What is the reason that the protease was not adsorbed to the starch?

ENDNOTES

1. Schwimmer, S., and A. K. Balls. 1949. Isolation and properties of crystalline α-amylase from germinated barley. *J. Biol. Chem.* **179**:1063-1084.

2. Ostrove, S. 1990. Affinity chromatography: general methods, p. 357-371. In: *Methods in Enzymology*, Vol. 182. M. P. Deutscher (ed.). Academic Press. New York.

3. Englard, S., and S. Seifter. 1990. Precipitation techniques, pp. 285-300. In: *Methods in Enzymology*, vol. 182. Deutscher, M. P. (ed.). Academic Press. New York.

NOTES

ION-EXCHANGE CHROMATOGRAPHY OF AMYLASE

One Class Period

Goals

You will learn how to construct an ion-exchange column and use it to purify amylase.

INTRODUCTION

Proteins contain both positively and negatively charged groups. The positively charged groups are protonated lysine and arginine residues, and the negatively charged groups are ionized glutamate and aspartate residues. As a consequence, proteins have a net charge that is dependent upon the pH. Because proteins have ionizable groups on their surface, they can be separated by ion-exchange chromatography. A column is packed with a stationary phase containing ionizable functional groups. An anion exchanger has fixed positive charges and displaceable negative counterions, whereas a cation exchanger has fixed negative charges and displaceable positive counterions. Any volume of sample can be applied to the column provided that the total amount of protein added does not exceed the binding capacity of the resin. It is even possible to concentrate proteins using ion exchange. A large volume of protein can be applied and eluted in a much smaller volume at the appropriate salt concentration. The length of the column is not important for ion-exchange chromatography. However, the total amount of

resin is important and must be sufficient to bind all the protein. Ultimately, this is determined by experimentation, although the manufacturer does provide information regarding the milliequivalents of ion that can be exchanged per dry gram or ml of resin bed. For anion exchangers, this would be milliequivalents of Cl^- (if the resin is supplied in the chloride form), whereas for cation exchangers it would be milliequivalents of H^+ (for resins supplied in the hydrogen form). You will be using an anion exchanger, that is., DEAE DE52 (DEAE cellulose). DEAE is diethylaminoethyl, which is a weak base and will have a net positive charge when ionized. It will therefore exchange negative charges (anions). Protein with negative charges will displace the negative counterions on the anion exchanger and bind to the stationary phase (resin). Positively charged proteins will not bind to the resin. The bound proteins can then be eluted with a high salt concentration (e.g., KCl), because the negative charges in the salt will displace the protein. You will bind the protein to the resin and elute with a gradient of KCl in Tris buffer. If a gradient maker is not available, the enzyme can be eluted using a step gradient. It should elute between 200 and 300 mM KCl.

MATERIALS

Supplies and equipment
- 30°C water bath
- side-arm flask for degassing slurry

For chromatography and assay each team should have the following (see Fig. 10.1):
- 10 cm x 1.5 cm column
- gradient maker
- magnetic stirrer
- stir bar
- ring stand and clamps
- glass beads, about 400 to 600 μm (Sigma Chemical Co. St. Louis, Mo.)
- 25 13 x 100 mm test tubes in a test tube rack,
- 50 16 x 150 mm test tubes
- 10 ml of swollen, equilibrated DE 52 to which 10 ml of 10 mM Tris, pH 8, containing 5 mM KCl has been added.

Solutions for equilibrating DE-52 (per team)
20 mM Tris, pH 8.0, 180 ml
10 mM Tris, pH 8.0, 60 ml

Solutions for gradient maker (per team)
Tris, 10 mM, pH 8, 5 mM KCl, 40 ml

Tris, 10 mM, pH 8, 300 mM KCl, 15 ml

Solutions for step elution (per team)
Tris, 10 mM, pH 8 containing:
5 mM KCL, 10 ml
50 mM KCl, 10 ml
100 mM KCl, 10 ml
200 mM KCl, 10 ml
300 mM KCl, 10 ml

Solutions for amylase assay (per team)
KI (0.2% I_2 + 0.3% KI), 30 ml

Potassium phosphate buffer, 0.1 M pH 6.5, 130 ml. (See Appendix C)

2% Soluble starch in 0.1 M potassium phosphate buffer, pH 6.5; autoclave into solution or steam for about 10 min, 3 ml

Enzyme (per team)
Each team should receive approximately 3 ml of dialyzed supernate from a stationary-phase culture of *Bacillus subtilis* (ATCC 35854). Grow the cells to stationary phase shaking at 30°C in nutrient broth containing 0.2% soluble starch. (See Exp. 8.) For example, inoculate 50 ml of medium in a 250 ml Erlenmyer flask with a loop of a plate culture and grow for 22 to 24 h. Centrifuge the cells at 8000 x g for 20 min. Dialyze the supernate overnight against 10 mM Tris buffer, pH 8.0, containing 5 mM KCl. Use 1 liter of buffer per 50 ml of supernate and change the buffer twice.

PROCEDURE

Equilibrating the DE-52 (per team)
1. Weigh out 4 g of DE-52 and slowly add it to 30 ml of 20 mM Tris, pH 8.0, with stirring.

2. Let sit in refrigerator overnight.

3. Pour off liquid.

4. Add fresh buffer and repeat 4 times.

5. Add fresh buffer and measure pH. If neces-

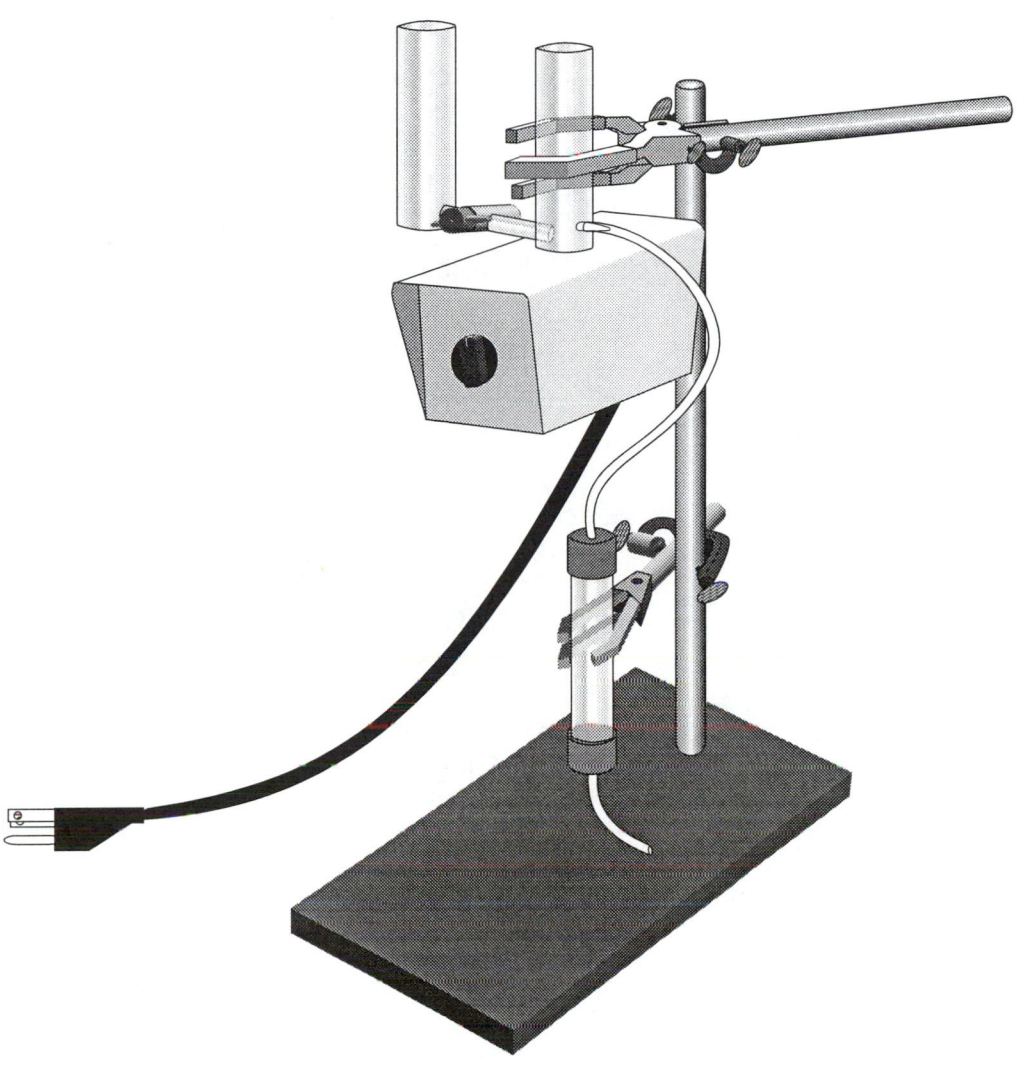

Fig. 10.1 Chromatographic column and gradient maker.

sary, adjust to pH 8.0 with HCl or NaOH.

6. Pour off liquid and add 30 ml of 10 mM Tris, pH 8.0, Allow to settle, pour off liquid and add fresh 10 mM Tris, pH 8.0.

7. Before use, bring to room temperature and degas for 10 min. To degas, pour slurry (10 ml buffer plus 10 ml of slurry) into a side-arm flask and place under vacuum using sink aspirator, shaking occasionally. the slurries for all of the teams can be combined and degassed together.

Pouring the column

1. Place about 2 mm of glass beads in the column. The beads cover the sintered glass bottom in the column so that the resin will not clog the column.

2. Clamp the exit tubing and pour the slurry into the column, being careful not to disturb the beads. The slurry volume should be about 20 ml (i.e.,10 ml resin and 10 ml of buffer, 10 mM Tris, pH 8.0, 5mM KCl).

3. Unclamp the exit tubing and allow the liquid to drain out. The resin will settle. Wash with 5 ml of the buffer containing 5 mM KCl. When the liquid comes to about 1 cm of the gel bed, clamp the tubing.

Eluting with gradient maker

1. Have ready 24 13 x 100 mm test tubes in a test tube rack. Mark the tubes showing a volume of 1 ml and label the tubes 1-24. Have an ice bucket nearby.

2. Close the valve between the two mixing chambers in the buffer reservoir and clamp the tubing leading from the reservoir to the column.

3. Add 12 ml of buffer containing 5 mM KCl to the reservoir leading to the column. This reservoir should also have a small magnet and be situated on a magnetic stirrer. In order to prevent an air bubble from forming between the two reservoirs, open the valve between the two reservoirs until a little bit of the 5 mM KCl solution passes into the empty reservoir. Close the valve. Remove the small volume of solution and return it to the first reservoir. Add 10 ml of buffer containing 300 mM KCl to the empty reservoir. Start the magnetic stirrer.

4. Add 2 ml of sample to the column very carefully so as not to disturb the bed.

5. Open the exit tubing and allow the sample to drain into the bed. Wash the column with 5 ml of buffer containing 5 mM KCl. This will remove positively charged proteins. Close the exit tubing.

6. Carefully add about 1.5 ml of buffer contining 5 mM KCl to the column. The buffer should be about 1 to 2 cm above the resin. Connect the gradient maker to the column and unclamp the exit tubing so that buffer flows into the column and begin collecting 1 ml fractions. When buffer begins to flow into the colmumn, open the valve between the reservoirs in the gradient maker. It is important not to open the valve between the reservoirs until buffer actually begins to flow into the column. At this point there should be about 0.5 cm to 1 cm of buffer on the top of the resin. If there is

too little buffer, you can add some. However, the volume above the resin should be small to prevent mixing of the buffer. As each fraction is collected, place the tube in the ice.

Eluting with a step gradient

If a gradient maker is not used, then the column can be eluted with a step gradient. After adding the sample and washing the column with 5 ml of buffer containing 5 mM KCl, the following procedure can be used:

1. Add 10 ml of buffer containing 50 mM KCl and collect 5 2 ml fractions.

2. When the meniscus of the buffer reaches the bed, add 10 ml of buffer with 100 mM KCl and collect 5 2 ml fractions.

3. Repeat with 200 and 300 mM KCl.

Assay for amylase

1. Add 5ml of 0.1M potassium phosphate buffer, pH 6.5 to 16 x 150 mm test tubes. You should have a test tube for every fraction collected, one tube for the unfractionated enzyme (labeled c, and 3 tubes that should be labeled 0. Number the rest of the tubes acccording to the fraction number.

2. Add 75 μl of 2% starch to each of the tubes and mix.

3. Add 1 ml of the iodine solution to 16 x 150 mm test tubes. There should be a test tube for every test tube that has buffer and starch and they should be labeled in the same way.

4. Remove 2 ml of the starch-buffer solution from each of the 3 test tubes labeled 0 and add to each of 3 tubes containing the iodine also labeled 0. Add 7 ml of water and mix. These tubes are the 0 time controls. They will be read and the average of their absorbance readings will be used.

5. To each of the remaining tubes add 200 μl of one of the enzyme fractions and incubate for 30 min at 30°C. Add contents of each to an iodine-containing tube and mix.

6. Read at 550 nm. Subtract each reading from the average of the 0 min readings and express the data as $A_{550}/30\text{min}$.

PREPARATION OF DATA

Subtract the reading at 30 min from the average of the three 0 time readings and plot this difference against fraction number.

QUESTIONS

1. Why was it important to dialyze the supernate against buffer at pH 8 and 5 mM KCl?

2. Anion exchangers can be used to concentrate proteins. Explain why this is the case.

3. Why is it important that the volume of eluting buffer on top of the resin be kept small when running the gradient?

REFERENCES

Rossomando, E. F. 1990. Ion-exchange chromatography,
  Vol. 182. p. 309-316. In: *Methods in Enzymology,*
  Deutscher, M. P. (ed.). Academic Press.

NOTES

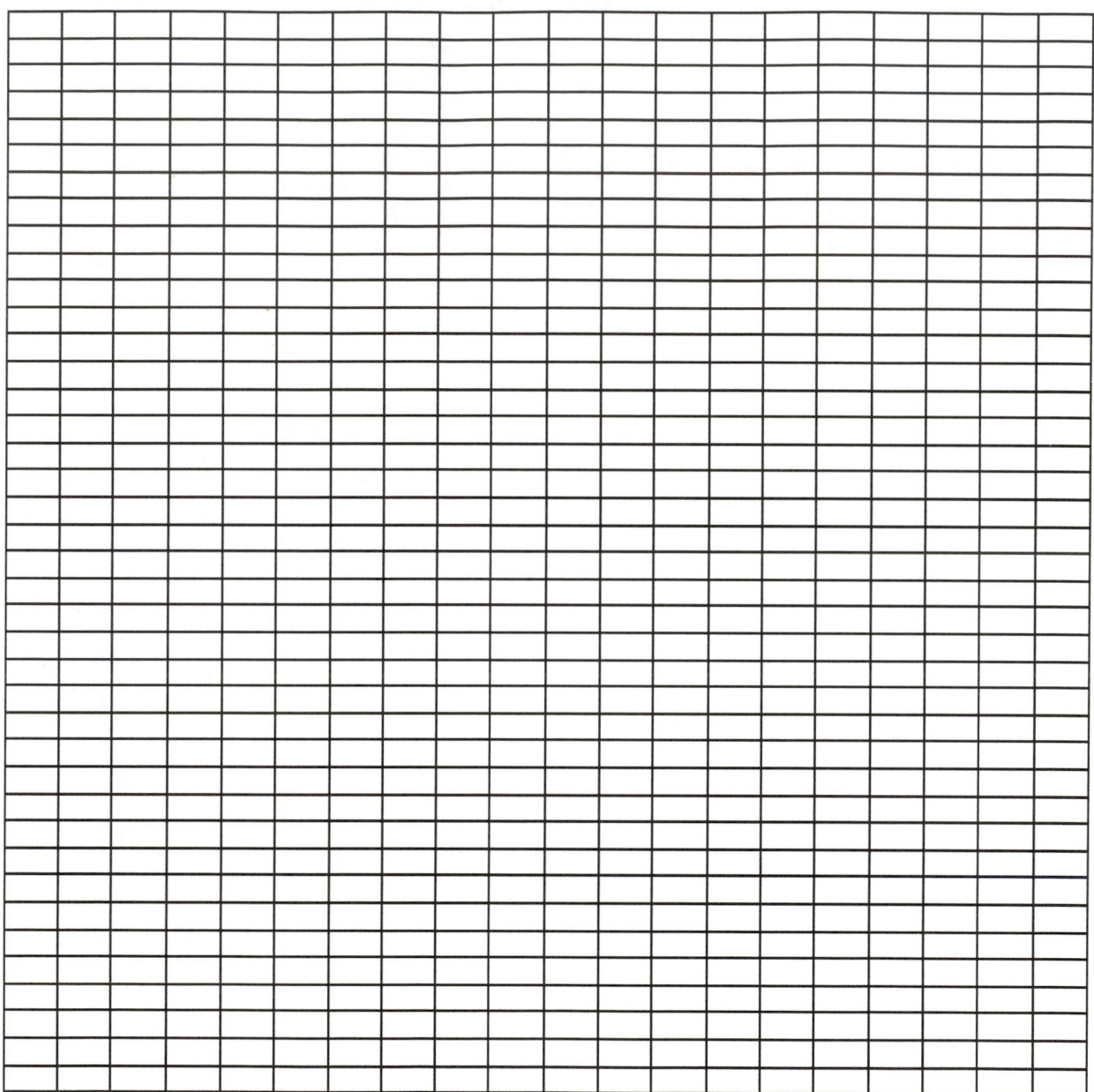

Experiment 10. Ion-Exchange Chromatography of Amylase

INDUCTION OF ALKALINE PHOSPHATASE AND THE DETERMINATION OF ITS CELLULAR LOCATION IN *E. COLI*

One Class Period

Goals

The goals of this experiment are to teach you (1) how to form spheroplasts of gram-negative bacteria, (2) how to assay for alkaline phosphatase, and (3) how to learn whether an enzyme is located in the periplasm of gram-negative bacteria. (For students who wish to determine the concentrations of inorganic phosphate that limit the growth yields of *E. coli*, or to examine the enzyme via electrophoresis, see Independent Projects in Appendix H.)

INTRODUCTION

Phosphatases are enzymes that catalyze the hydrolysis of phosphate esters to produce inorganic phosphate. The reaction is:

$$R-O-\overset{\overset{O}{\|}}{\underset{\underset{O^-}{|}}{P}}-O^- + H_2O \longrightarrow R-OH + HPO_4^{2-}$$

or

$$R-O-\textcircled{P} + H_2O \longrightarrow R-OH + P_i$$

Some phosphatases such as fructose-1,6-bisphosphatase which is studied in Experiment 13 are specific for the substrate that is attacked. The acid and alkaline phosphatases are not specific with regards to the substrate and attack a variety of different organic phosphate esters. The acid phosphatases have pH optima below 7, and the alkaline phosphatases have pH optima above 7. For example, the alkaline phosphatase from *E. coli* has a pH optimum of about 8. It is located in the periplasm (the area between the inner and outer membranes) and serves to provide inorganic phosphate for the bacteria when inorganic phosphate in the medium becomes limiting for growth. See Fig. 11.1 for a schematic drawing of the envelope of *E. coli*, including the periplasmic region. The proteins labelled Pr in Fig.11.1 represent the periplasmic proteins. It is necessary to remove the phosphate from the organic phosphates in the periplasm rather than the cytoplasm because, whereas *E. coli* has a transport system for inorganic phosphate, it generally cannot transport organic phosphates into the cell. The presence of excess inorganic phosphate in the external medium represses the synthesis of alkaline phosphatase by *E. coli*. Thus the cells must be grown under inorganic phosphate limitation conditions in order to stimulate the synthesis of alkaline phosphatase. One way to do this is to grow *E. coli* until the inorganic phosphate has been depleted from the medium and

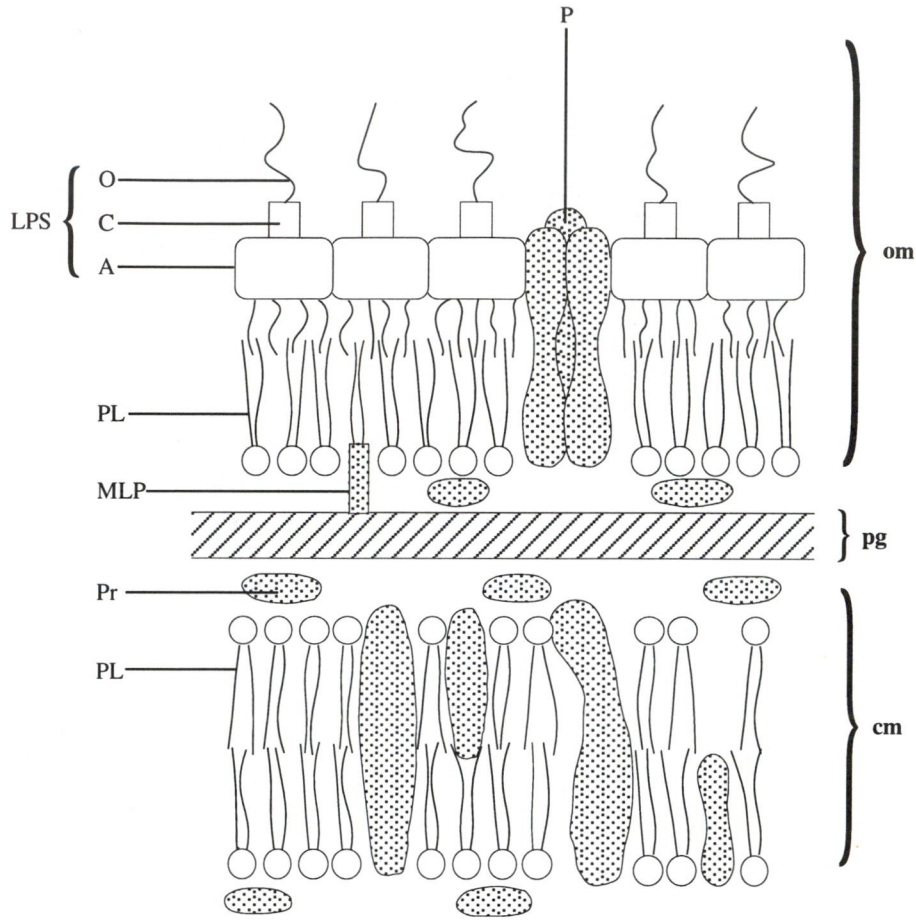

Fig. 11.1 Schematic drawing of the gram-negative envelope showing the periplasmic region. The outer membrane consists of lipopolysaccharide, phospholipid, and proteins, most of which are porins. Underneath the outer membrane is the peptidoglycan layer, which is noncovalently bonded to the outer membrane via murein lipoproteins, themselves covalently attached to the peptidoglycan. The cell membrane is composed of phospholipid and protein. The area between the outer membrane and the cell membrane is called the periplasm. The wavy lines are fatty acid residues which anchor the phospholipids and lipid A into the membrane. Abbrev. LPS, lipopolysaccharide; O, oligosaccharide; C, core; A, lipid A; P, porin; PL, phospholipid; MLP, murein lipoprotein; pg, peptidoglycan; Pr, periplasmic proteins; om, outer membrane; cm, cell membrane. (From White, D.,1995. *The Physiology and Biochemistry of Prokaryotes.* Oxford University Press, New York.)

then assay for alkaline phosphatase.

E. coli will be grown in limiting phosphate (0.1 mM) until the population of cells have been in stationary phase for at least 2 h, by which time the inorganic phosphate has been depleted. These cells will be assayed for alkaline phosphatase. You will also assay the enzyme in cells that were grown in excess phosphate (0.5 mM).

The stationary phase cells will be harvested and treated with lysozyme and EDTA. The combination of lysozyme and EDTA will release the periplasmic proteins, including alkaline phosphatase, into the medium. Studies have shown that cytoplasmic proteins are not released under these conditions.

When the peptidoglycan is destroyed by the lysozyme, the cell loses its rigid cell wall. Under such circumstances it would be expected that water would rush into the hyperosmotic cell and lyse it since the rigid cell wall has been destroyed and now nothing prevents the water from entering. However, the lysozyme treatment is done in the presence of 20% sucrose, which is approximately isoosmolar with the cell interior so that water does not rush in. The EDTA removes divalent cations which hold the lipopolysaccharide (LPS) to the outer envelope. Consequently, a major portion of the LPS is lost. This presumably is a factor in the release of the periplasmic enzymes and in allowing the lysozyme to reach the peptidoglycan. You can expect that the cells grown in excess phosphate will become phase-light upon treatment with lysozyme and EDTA. Some of them may round up, but most will remain rod-shaped, albeit greatly expanded rods since the peptidoglycan is destroyed by the lysozyme. The phosphate-limited cells will be much shorter than the cells grown in excess phos-

phate. You may not notice much of a change upon treatment with lysozyme and EDTA, nevertheless the alkaline phosphatase is released.

The assay for alkaline phosphatase relies on the hydrolysis of *p*-nitrophenylphosphate (PNPP). This is an organic phosphate that when hydrolyzed yields inorganic phosphate and nitrophenol. The nitrophenol is yellow and can be measured at 420 nm.

MATERIALS

Supplies and equipment
- refrigerated centrifuge and tubes for centrifuging the cells
- spectrophotometer or colorimeter at 420 and 590 nm
- 30^0C water bath
- phase-contrast microscopes, 40x objectives
- microscope slides and coverslips
- sterile 16 x 150 mm test tubes
- magnetic stirrer and stirring bars

Solutions

Potassium phosphate. 0.05 M K_2HPO_4

20% Sterile glucose. Will dilute this 50 fold into growth medium after autoclaving.

Growth medium. SMB* containing NH_4Cl as the nitrogen source and 0.4% glucose (added separately after autoclaving) and 0.01 M morpholinepropanesulfonic acid (MOPS) buffer, pH 7.4, instead of phosphate buffer. (See Appendix C.) It is convenient to make a 0.5 M buffer and dilute it when making the media. Add 0.05 M K_2HPO_4 diluting it to 0.1 mM (phosphate-limiting) or 0.5 mM (phosphate excess). Autoclave everything together except the glucose which can be added after autoclaving as a sterile 20% solution to a final concentration of 0.4%.

Tris-HCl buffer, pH 8.0. (Make a 1 M solution and dilute it appropriately.) 0.01 M for washing cells (will need approximately 20 ml for each 50 ml culture), and 0.033 M containing 20% sucrose for spheroplasting the cells. Will need approximately 40 ml for two 50 ml cultures. Will need approximately 25 ml of the 1.0 M Tris buffer for each team for the PNPP.

EDTA, 10 mM, pH 8. Make 100 ml.

Lysozyme. crystalline egg white lysozyme, 1 mg/ml in 0.033 M Tris-HCL, pH 8. Make 5 ml. It can be kept in the refrigerator for several days or stored frozen for longer periods.

p-Nitrophenylphosphate (PNPP) 1.0 mM in 1M Tris buffer, pH 8. Each team will require approximately 25 ml. Make fresh on day of use.

Cells

50 ml of *E. coli* B grown at least 2 h into stationary phase in a 250 ml Erlenmyer or Klett flask shaking at 37°C. Grow the inoculum in regular SMB*, pH 7.4, containing 0.4% glucose (added separately after autoclaving). To grow the inoculum transfer a loop of cells from a slant or plate into 50 ml of the SMB* and incubate on a shaker at 37°C overnight. See Appendix C for a description of how to make SMB*. Inoculate the 50 ml of growth medium for class with 50 µl of the inoculum and grow 16 to 20 h. The Klett reading (#66 filter) will be approximately 100 for the phosphate-limited culture and about 230 for the phosphate-excess culture. Each team of students should have two cultures, one of which

was grown in limiting phosphate (0.1 mM), and the other in excess phosphate (0.5 mM). (See "growth medium".) The 50 ml of culture should provide approximately 10 ml of enzyme, and each team will use about 1 ml. Thus several teams can share the same culture. The culture can be used at any time during the day or, if preferred, placed in the refrigerator and used the next day.

PROCEDURE

Preparation of enzyme

(Making spheroplasts)

1. Measure the absorbance of the cell suspension at 590 nm. You will need this value when you resuspend the cells in step 2. Centrifuge the cells at 12000 x g for 10 min and wash twice with 10 ml of 0.01M Tris-HCl, pH 8.0.

2. Using the absorbance measurement of step 1, concentrate the cells to about an absorbance of 3 at 590 nm by resuspending with Tris buffer (0.033M), pH 8.0, containing 20% sucrose. You can easily calculate the volume of cell suspension with which you must start in order to obtain an absorbance of 3. Use the relationship $A_1 V_1 = A_2 V_2$, where A is the absorbance and V is the volume. The cells will be concentrated around 2.5-fold.

4. Stir the cell suspension very slowly at room temperature with a magnetic stirring bar in a beaker and add 10 mM EDTA, pH 9, to a final concentration of 0.1 mM, and lysozyme to a final concentration of 10 µg/ml.

5. Examine periodically to monitor spheroplast formation using the microscope. The phos-

phate-limited culture may not appear to be spheroplasting, but the lysozyme-EDTA treatment nevetheless releases alkaline phosphatase into the medium.

6. Wait 20 min and then centrifuge the suspension at 8500 x g for 15 min.

7. Remove the supernate and assay for activity.

Assay for alkaline phosphatase

1. Add enzyme at a 1:30 dilution to PNPP solution in test tube or colorimeter tube and read the absorbance at 420 nm. This is the 0 time reading.

2. Incubate the tubes in a $30^\circ C$ water bath and at 10, 20, and 40 min take absorbancy readings.

PREPARATION OF DATA

Plot absorbancy readings against time on linear graph paper. If you wish, convert the absorbancy readings to amount of PNPP hydrolyzed. The molar extinction coefficient for *p*-nitrophenol is 1.32×10^4. (See problem #8 in Appendix F.) Define an enzyme unit as a change of $0.01\ A_{420}$ unit per minute and note the number of units of enzyme per ml of the enzyme preparation.

QUESTIONS

1. Excess phosphate has been reported to repress the synthesis of alkaline phosphatase in *E. coli*. Are your data in agreement with the published results?
2. What experiment might you perform to support the conclusion that the phosphatase was indeed released from the periplasm and not from the cytoplasm?

REFERENCES

Malamy, M., and B. L. Horecker. 1966. Alkaline phosphatase (crystalline), pp. 639-642. In: *Methods in Enzymology*, Vol. IX. Wood, W. A. (ed.). Academic Press, New York.

NOTES

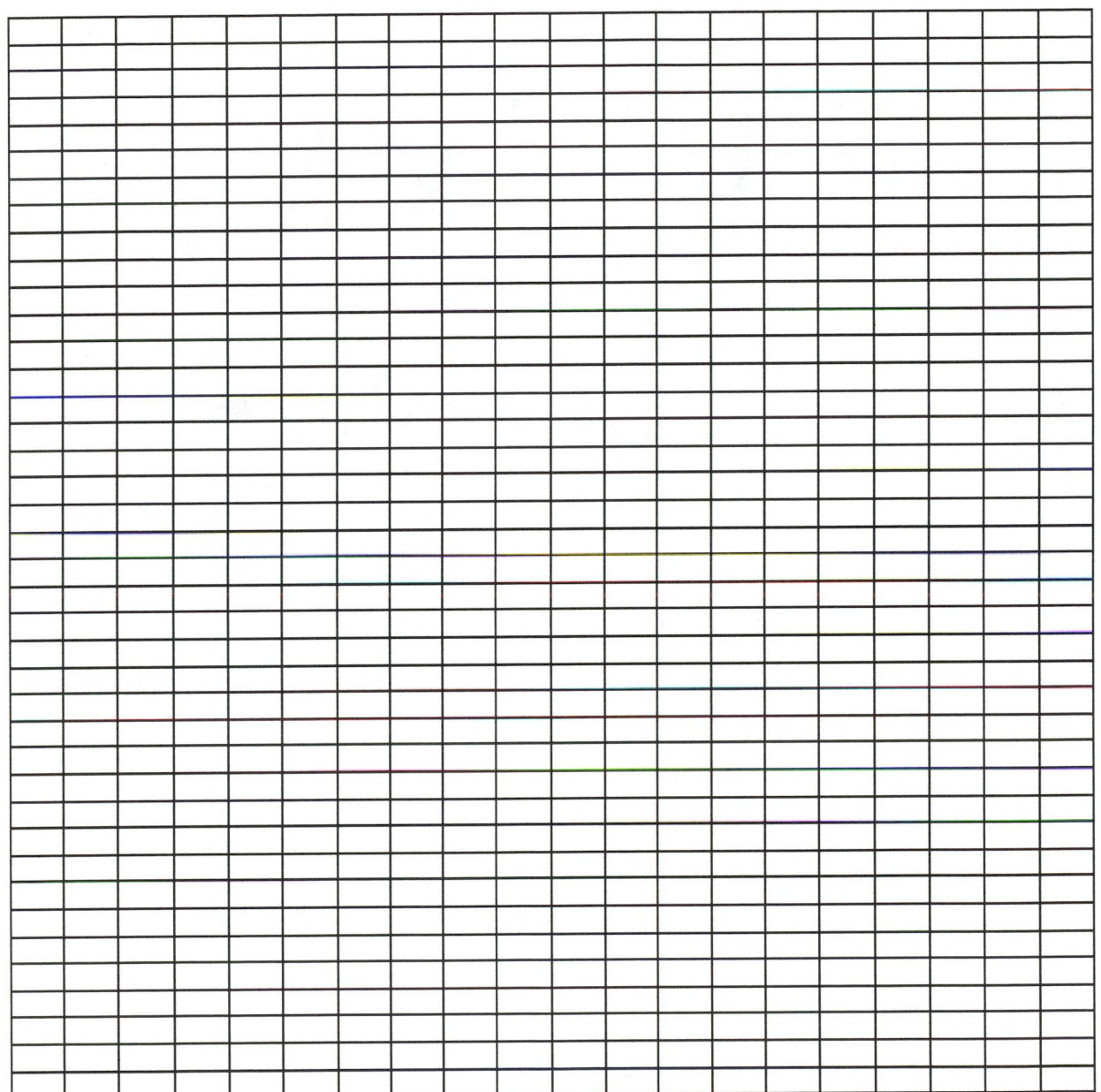

ASSAY OF THREONINE DEAMINASE: DETERMINATION OF K_M AND V_{max}

Two Class Periods

Goals

The goals of this experiment are: (1) to teach you how to assay the enzyme threonine deaminase, and (2) to show you how to determine the K_m and V_{max} of an enzyme.

INTRODUCTION

Threonine deaminase catalyzes the following reaction (Fig.12.1):

$$\text{L-threonine} \longrightarrow \alpha\text{-ketobutyrate} + NH_4^+$$

The enzyme assay relies upon the fact that 2,4-dinitrophenylhydrazine reacts with the keto group of α-ketobutyrate to form a compound (hydrazone) whose absorbance is measured with a spectrophotometer. Many bacteria make two threonine deaminases, a biodegradative enzyme and a biosynthetic enzyme. The biodegradative en-

zyme is inducible. It is made when the cells are growing on rich media where L-threonine in excess can serve as a source of carbon, and glucose and oxygen are lacking. The threonine is deaminated to α-ketobutyrate, which in turn is converted to propionyl-CoA. The propionyl-CoA can be metabolized to propionate with the synthesis of an ATP. The biodegradative enzyme is activated by AMP, which can be thought of as a signal that the ATP levels are low. The biosynthetic threonine deaminase is made in high amounts when *E. coli* is grown on a minimal medium with glucose as the source of carbon. The α-ketobutyrate that is formed is a precursor to isoleucine, which is a feedback inhibitor of

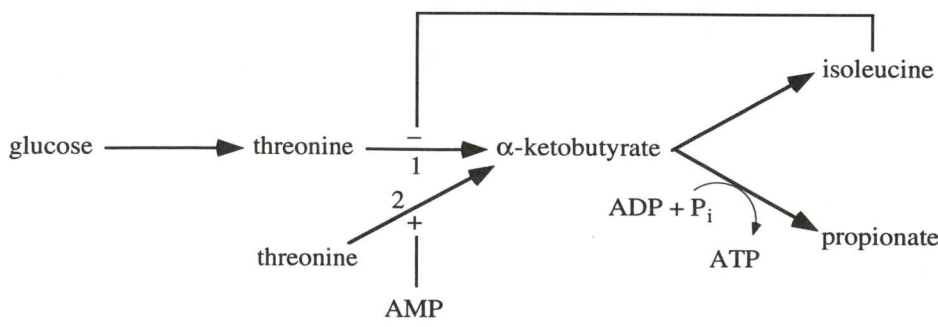

Fig.12.1 Regulation of threonine deaminase activity: (1) biosynthetic enzyme; (2) degradative enzyme.

the activity of the biosynthetic threonine deaminase.

Enzymes that follow Michealis-Menten kinetics show a relationship between the velocity (v) and substrate concentration (S) according to:

$$v = V_{max}S/(K_m + S) \qquad (1)$$

This predicts a curve showing saturation kinetics when v is plotted against S. The velocity approaches a maximum called V_{max} as the enzyme becomes saturated with substrate. When v is set equal to $(1/2) V_{max}$ then solving for S reveals that $S = K_m$. In other words, the K_m is the substrate concentration that gives $(1/2) V_{max}$. Because V_{max} cannot be determined accurately when plotting v versus S, the Lineweaver-Burke plot is often used. In the Lineweaver-Burke plot $1/v$ is plotted against $1/S$ to give a straight line with a slope of K_m/V_{max} that intersects the y and x axes. The equation for Lineweaver-Burke kinetics is derived from the Michealis-Menten equation by taking its reciprocal:

$$1/v = 1/V_{max} + (K_m/V_{max})(1/S) \quad (2)$$

You can see that if you set $1/S$ to 0, then $1/v = 1/V_{max}$. That is to say, $1/V_{max}$ is where the line intercepts the y axis. On the other hand, if you set $1/v$ to 0 then $1/S = -1/K_m$, that is, $-1/K_m$ is where the line intercepts the x axis. The Lineweaver-Burke plot is drawn in Fig. 12.2.

MATERIALS

Equipment

- Spectrophotometer for 416 and 660 nm readings

Solutions for making extracts (per class)

Washing buffer, 200 ml. 0.05M potassium phosphate buffer, pH 7.4, (Appendix C).

Breakage buffer, 30 ml. 0.05M potassium phosphate buffer, pH 7.4, 0.01M AMP, 2 mM mercaptoethanol (7 µl per 50 ml buffer).

Solutions for measuring α-ketobutyrate (per team)

A-1, 12 ml.
  2.4 ml 0.5 M potassium phosphate buffer, pH 7.4.
  1.2 ml 0.1 M AMP dissolved in 0.5 M potassium phosphate buffer, pH 7.4.
  2.4 ml 1.0 M DL-threonine dissolved in distilled water.
  6.0 ml distilled water.
0.01 M α-Ketobutyrate , sodium salt, 0.5 ml.

Solutions for assaying enzyme (per team)

A-2 (A-1 without threonine), 34 ml
  8.4 ml 0.5 M potassium phosphate buffer, pH 7.4
  4.2 ml 0.1 M AMP dissolved in 0.5 M potassium phosphate buffer
  21 ml distilled water

Note: This amount of A-2 provides an extra 8 ml per team, in case the assay to determine a linear rate must be repeated.

DL-threonine. 1.0 M, 4 ml; 0.5 M, 0.33 ml.

0.1% 2,4-dinitrophenylhydrazine in 2N HCl. 8.4 ml. (Shake overnight at 37°C to dissolve.)

1 N HCl, 42 ml.

2 N NaOH, 84 ml.

Solutions for Lowry assay: (per team)
(See Appendix C.)

Lowry solution C, 60 ml.

Folin-phenol/water, 7 ml.

1.00 mg/ml bovine serum albumin, about 0.6 ml.

Growth of cells

About 2 liters of *E. coli B* are grown at 37°C, without shaking, for 16 h until they reach stationary phase. One may inoculate 900 ml of medium in a 1 liter flask with a loop of cells from a plate. The growth yields are relatively low. Under these conditions the cells may grow to a turbidity of about 18 to 20 (Klett #66). (The A_{660} using a spectrophotometer is about 0.2 to 0.4 depending upon the spectrophotometer.) The growth medium consists of the following:

> 0.5% K_2HPO_4
> 2% Bactopeptone
> 0.4% DL-threonine
> 0.4% DL-serine
> pH 7.5

Making cell extracts (each team will need approximately 1 ml if the assay is done twice) (Prepared by instructor before class.)

1. Centrifuge cells at 8000 x g for 20 min and combine the pellets by washing with about 100 ml of washing buffer.

2. Suspend with 25 ml cold breakage buffer and break at 12000 psi in a cold French pressure cell.

3. Centrifuge extract at 12000 x g for 15 min. Remove supernatant fluid (crude extract) to chilled tube. The enzyme can be kept refrigerated at 4°C for at least 48 h. Do not freeze.

PROCEDURE - DAY 1

Determining the relationship between absorbance and α-ketobutyrate

1. Set up 11 test tubes each containing 1.0 ml of A-1.

2. Tubes 2-11 receive α-ketobutyrate for a standard curve with 5 points over a range of 5 to 50 μl. Do duplicate tubes. Tube 1 is used to zero the spectrophotometer.

2. Add 1.0 ml of 1.0N HCl to each tube.

4. Add 0.2 ml of dinitrophenylhydrazine to each tube. Mix and incubate 10 min at room temperature.

5. Add 2 ml of 2.0N NaOH, mix, and read absorbance at 416 nm.

Plot absorbance on ordinate and μmoles of α–ketobutyrate on abscissa

Enzyme assay to determine the amount of enzyme that produces a linear rate for at least 10 min

In this experiment you will incubate enzyme and saturating substrate for different lengths of time. The object is to find the amount of enzyme that yields linear kinetics for at least 10 min. You will need to know this in order to do the next experiment that will measure the K_m and V_{max}. Plot your data as absorbance change (ordinate) versus time (abscissa). Make certain that the stock enzyme preparation remains in ice when it is not being used.

Set up 9 test tubes. (one tube will be used to zero in the spectrophotometer and 8 will be used to run duplicates) and add:

1. 0.8 ml of A-2 (room temperature) to each tube;

2. 50 µl of enzyme to each tube;

3. 1.0 ml of 1.0 HCl to tube 1; this will be used to zero the spectrophotometer;

4. Start the reaction by adding 200 µl of 1.0 M DL-threonine to each tube;

5. Stop reaction at 5, 10, 15, and 20 min by adding 1 ml of 1.0N HCl. the reactions are run in duplicate. Mix.

6. After all the tubes have received HCl, add 200 µl of dinitrophenylhydrazine to all the tubes, mix, and let sit at room temperature for 10 min.

7. Add 2 ml of 2.0 N NaOH to each tube, mix, and read at 416 nm.

You may have to repeat the experiment with a different amount of enzyme to obtain a linear rate for at least 10 min.

Determining the K_m and V_{max}

Choose a time period and enzyme amount for which the reaction is linear. Use the results of the previous enzyme assay as a guide. For this experiment, the stock DL-threonine should be 0.5M rather than 1.0M.

Set up 23 test tubes, each containing the following (duplicate tubes except for the blank):

1. 0.8 ml of A-2;

2. 0, 2.5, 5, 10, 15, 20, 25, 30, 40, 80, 100 µl of 0.5 M DL threonine. One tube should have 2.5 µl of 0.5 M DL-threonine diluted 1:1.

3. Water to 1 ml.

The blank is the 0 time tube. It will receive no substrate and HCl is added before the enzyme.

4. Start the reaction by adding enzyme and incubate all the tubes for a set period of time over which the reaciton is still linear.

5. Add HCl and finish the assay in the usual way. Plot the number of µmoles/ml of DL-threonine on the abscissa and the velocity on the ordinate. Also plot the data as $1/v$ versus $1/S$. Calculate the K_M and V_{max} using the $1/v$ versus $1/S$ plot. The enzyme does not use D-threonine and you should take this into account when calculating the K_M.

PROCEDURE - DAY 2

Protein assay (Lowry)

Assay in duplicate 5, 10, 20, and 60 µl of sample. See Experiment 6 for instructions. Use the data to calculate the specific activity of the enzyme.

PREPARATION OF DATA

Determining the K_m and V_{max}

Using linear graph paper, plot $1/v$ vs $1/S$ on one graph paper and v vs S on another. Determine the K_M and V_{max} as shown in Fig. 12.2.

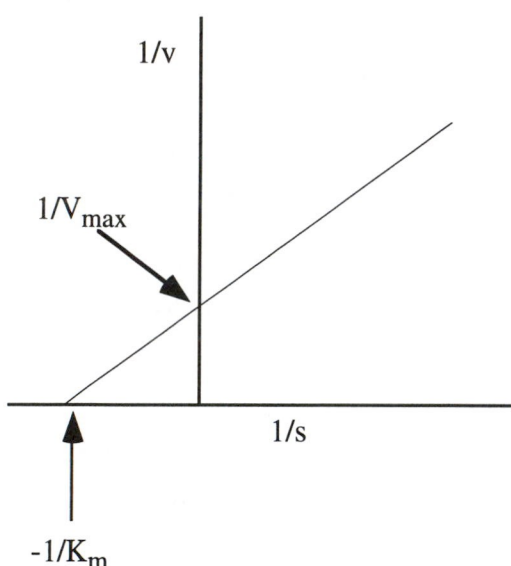

Fig. 12.2 Lineweaver-Burke plot.

Calculating the specific activity

Using the V_{max}, the standard curve of α-ketobutyrate versus absorbance determined on day 1 and the protein concentration, calculate the specific activity in terms of units of enzyme per μg of protein, where one unit is equal to one μmole of α-ketobutyrate formed per minute.

QUESTIONS

1. What are the definitions of K_M, V_{max}, and specific activity?

2. How do the K_M and V_{max} values that you determined compare with the class values? Are the values that you obtained reasonable? Explain.

REFERENCES

Dobrogosz, W. J. 1981. Enzymatic activity. pp. 365-392. In: *Manual of Methods for General Bacteriology*. Gerhardt, P.. Murray, R. G. E., Costilow, R.N., Nester, E. W., Wood, W. A., Krieg, N. R., and Phillips. G. B. (eds.). American Society for Microbiology. Washington, D.C.

NOTES

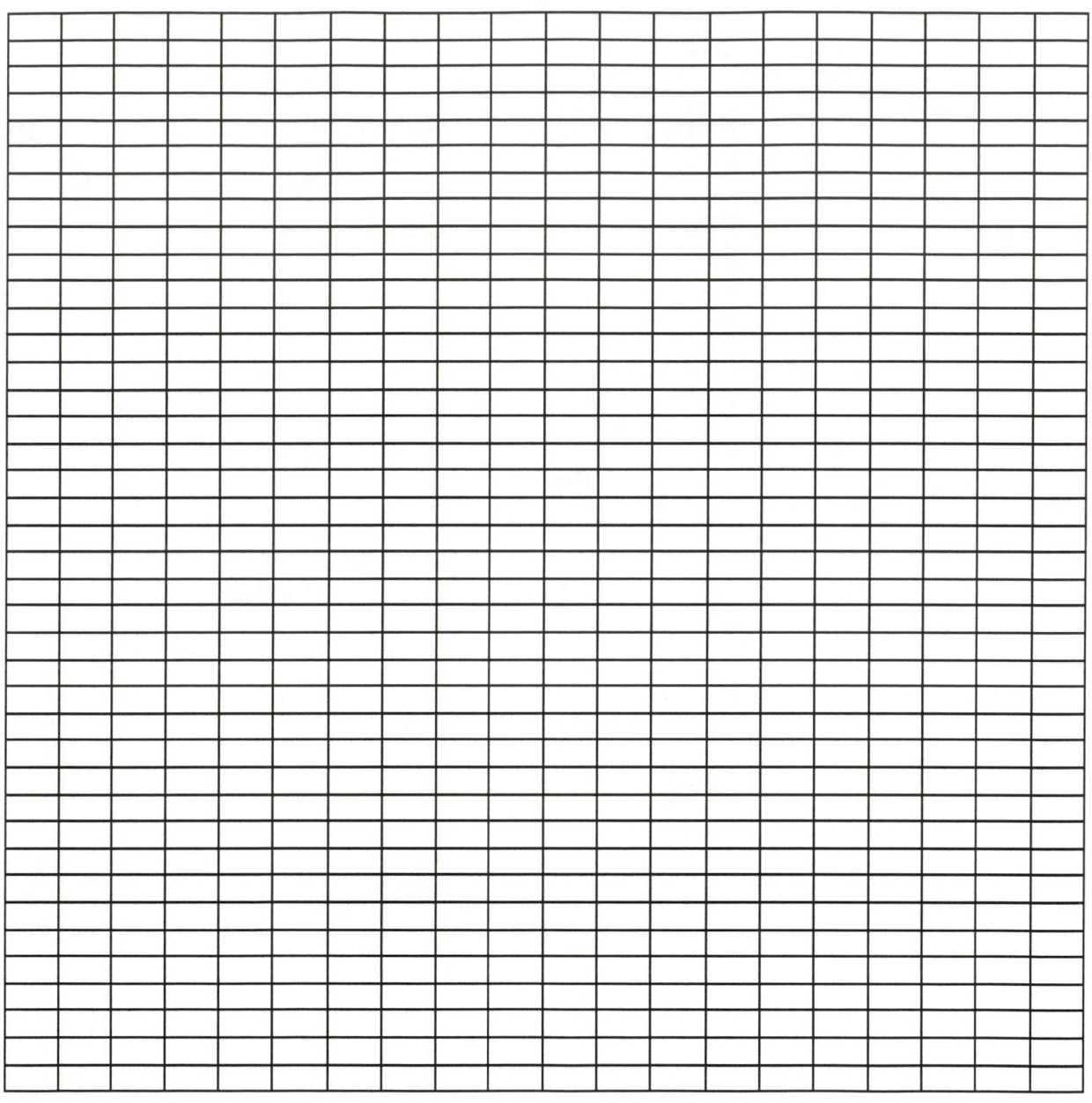

Experiment 12. Assay of Threonine Deaminase: Determination of K_M and V_{max}

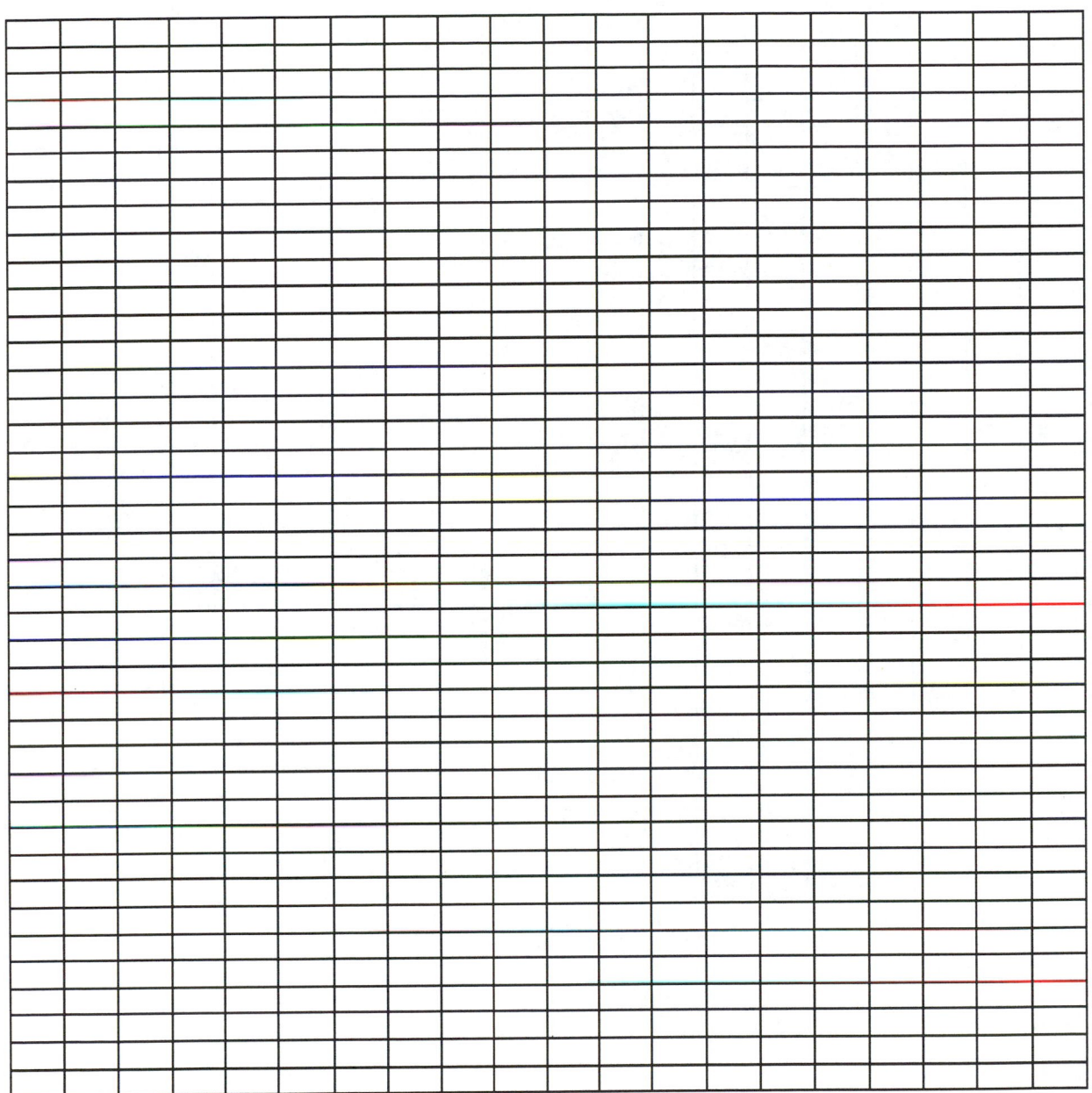

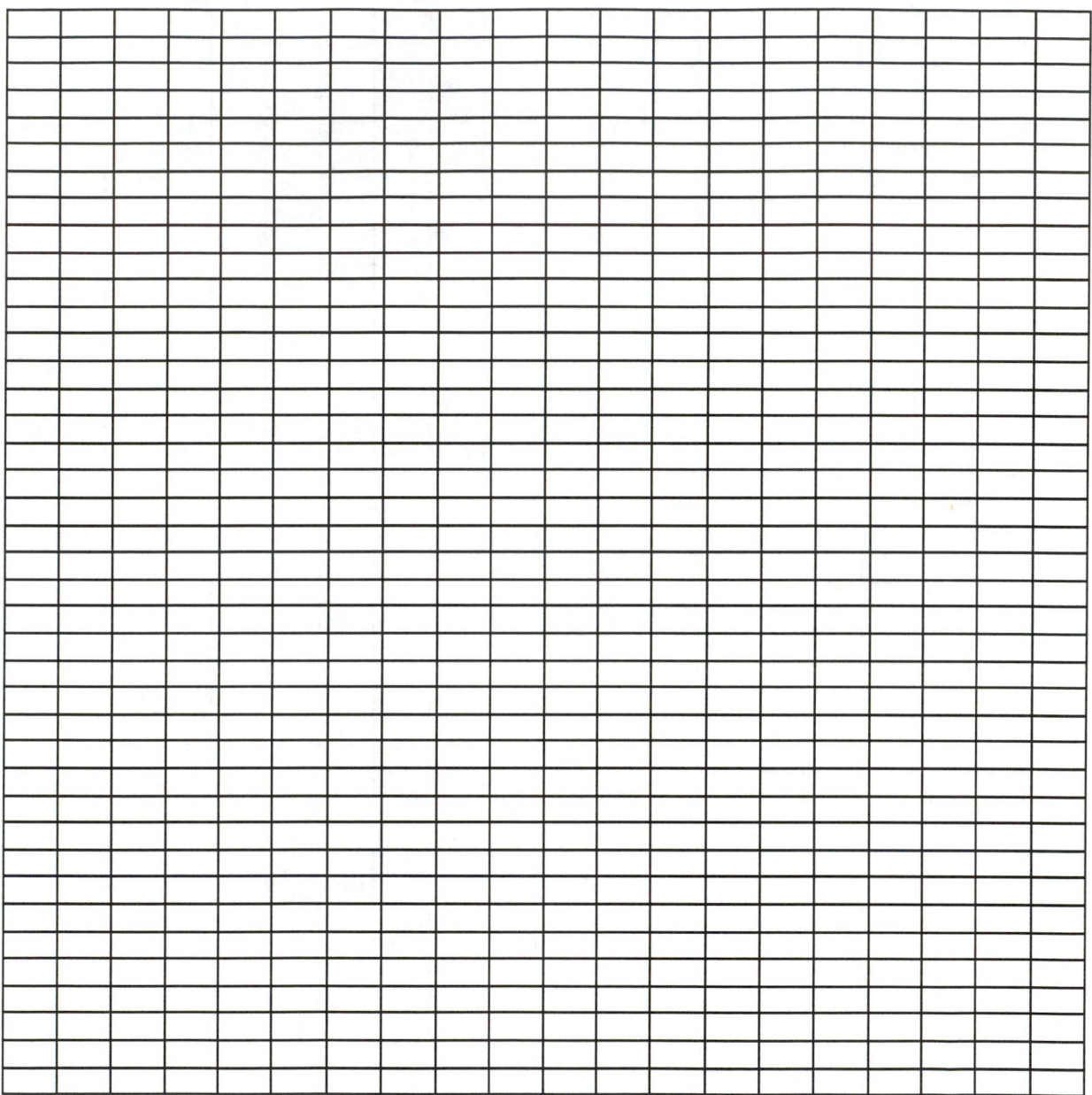

Experiment 12. Assay of Threonine Deaminase: Determination of K_M and V_{max}

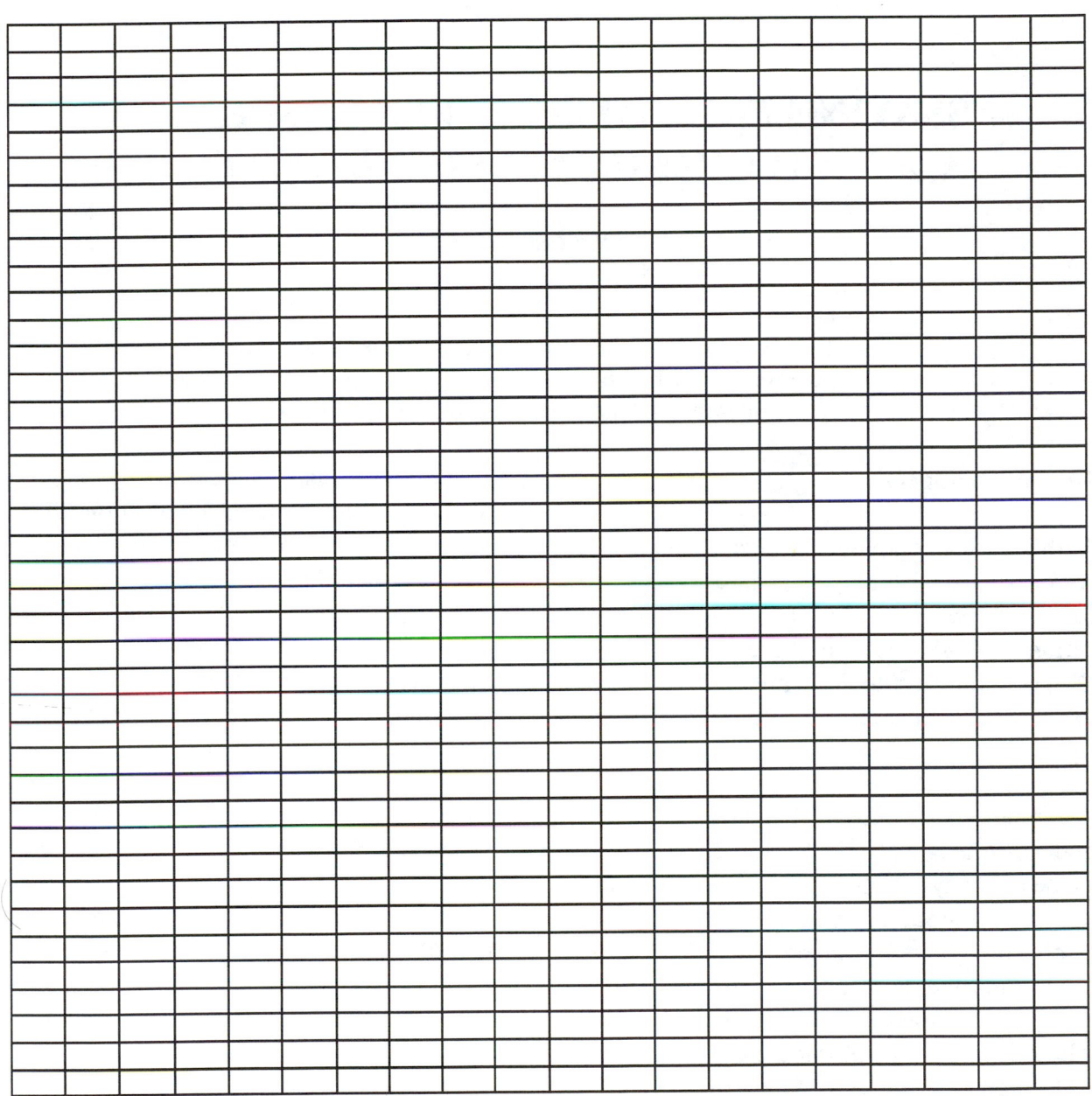

ASSAYING FRUCTOSE-1,6- BISPHOSPHATASE IN *SACCHAROMYCES CEREVISIAE*

One Class Period

Goals

The goals of the experiments are to (1) learn how to assay for fructose-1,6-bisphosphatase using coupled enzyme reactions, and (2) examine the inhibition of fructose-1,6-bisphosphatase by AMP.

INTRODUCTION

In this experiment you will determine the activity of fructose-1,6-bisphosphatase in *Saccharomyces cerevisiae* extracts and examine its inhibition by AMP. Refer to Appendix H for an independent project to determine the K_M and V_{max} as well as a further examination of the inhibition by AMP.

Examine Figure 13.1 which diagrams glycolysis and gluconeogenesis in *S. cerevisiae*. The part of the pathway which concerns us today includes enzymes 1 through 3 and enzyme 10. Enzyme 1 is phosphoglucose isomerase which reversibly transforms glucose-6-phosphate into fructose-6-phosphate. Enzyme 2 is phosphofructokinase, which catalyzes the physiologically irreversible ATP-dependent phosphorylation of fructose-6-phosphate to form fructose-1,6-bisphosphate. Enzyme 10 is fructose-1,6-bisphosphate phosphatase which catalyzes the physiologically irreversible hydrolysis of the C-1 phosphate from fructose-1,6-bisphosphate to form fructose-6-phosphate. This is the enzyme that you will assay here.

Fructose-1,6-bisphosphate phosphatase is assayed via coupled enzymatic reactions. The substrate, fructose-1,6-bisphosphate is cleaved to fructose-6-phosphate by the fructose-1,6-bisphosphatase (enzyme 10). The reaction mixture contains phosphoglucose isomerase (enzyme 1), glucose-6-phosphate dehydrogenase (enzyme 11), and NADP$^+$. The isomerase converts the fructose-6-phosphate to glucose-6-phosphate and the dehydrogenase catalyzes the reduction of NADP$^+$ by the glucose-6-phosphate producing NADPH. The reaction is measured by following the absorbance of NADPH at A_{340}. (The glucose-6-phosphate dehydrogenase is not part of glycolysis or gluconeogenesis. It is actually the entrance reaction into the pentose phosphate and Entner-Doudoroff pathways.)

The phosphatase operates only during gluconeogenesis and the fructokinase only during glycolysis, whereas the isomerase operates during both glycolysis and gluconeogenesis. The physiological activity of the phosphatase is believed to be controlled by AMP, which is an allosteric inhibitor in vitro. The inhibition by AMP is rationalized by assuming that when AMP levels are

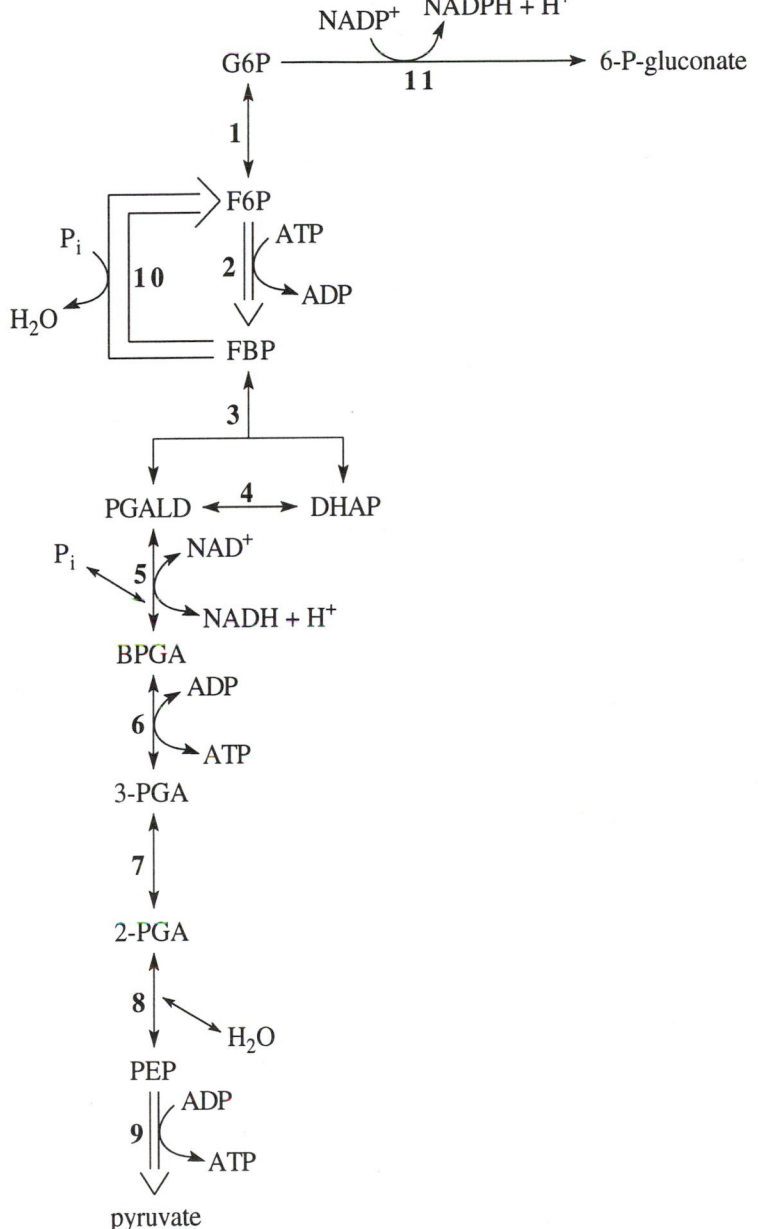

Fig. 13.1 Glycolysis and gluconeogenesis. Enzymes: (1) glucose isomerase; (2) phosphofructokinase; (3) fructose-1,6-bisphosphate aldolase; (4) triosephosphate isomerase; (5) triosephosphate dehydrogenase; (6) phosphoglycerate kinase: (7) mutase; (8) enolase; (9) pyruvate kinase; (10) fructose-1,6-bisphosphatase; (11) glucose-6-phosphate dehydrogenase.

high, ATP levels are low and the cell slows gluconeogenesis relative to glycolysis in order to synthesize more ATP.

MATERIALS

Solutions for cell extracts and dialysis
0.05 M potassium phosphate buffer, pH 7.4. 2 liters. (See Appendix C.)

Solutions for Lowry assay
See Experiment 4 and Appendix D.

Solutions for fructose-1,6-bisphosphatase
1 M imidazole buffer, adjusted to pH 7 with 5 N HCl, 100 ml. This is a 20x solution, which is used to make Solution A-1.

Solution A-1. 0.1 M KCl, 10 mM $MgCl_2$, 1 mM K_2EDTA, 50 mM imidazole buffer, pH 7. Can make 100 ml of a 10x stock solution. It is also used for Experiment #14.

Solution A-2. Add $NADP^+$ to a 1x Solution A-1 to a final concentration of 0.25 mM. Then add per ml of Solution A-1: 0.28 units of phosphoglucose isomerase (D-glucose-6-phosphate-ketoisomerase, EC 5.3.1.9) and 0.28 units of glucose-6-phosphate dehydrogenase (D-glucose-6-phosphate:NADP1-oxido-reductase, EC 1.1.1.49). The enzymes may be too concentrated as purchased. If necessary, add 1 ml of 1X solution A-1 to the enzymes in order to transfer the appropriate number of units to solution A-2. Solution A-2 can be stored at 4^0C for at least 36 h. Keep it cold until it is ready to be used. Then, let it come to room temperature just before use.

Each assay requires 5 ml. Each team may do as many as 3 assays.

1 mM fructose-1,6-bisphosphate in water. Each assay requires 0.4 ml. Therefore make 12 ml per team (3 assays).

50 mM AMP in water. Each team will be using approximately 0.2 ml.

Growth medium
glucose	10 g
$MgSO_4$	1 g
KH_2PO_4	2 g
peptone	6 g
yeast extract	4 g
distilled water	1 l

Growth of cells
S. cerevesiae is grown in 500 ml of media, shaking at 30^oC for 40 to 60 h. The inoculum is 0.5 ml of an overnight culture. The cells should enter stationary phase by 40 h. The enzyme increases in activity when the cells enter the stationary phase of growth. Harvest in the cold at approximately 8000 x g for 15 min.

PROCEDURE

Enzyme preparation
Wash cells in phosphate buffer (0.05 M potassium phosphate, pH 7.4) and resuspend with about 20 ml buffer and break in a French pressure cell. You may need to pass the material through the press 3 times to get good breakage. Centrifuge the broken cell suspension at 24000 x g for 20 min. Dialyze the supernate in the coldagainst 1 liter of phosphate buffer for 18 h, changing the buffer 2 times. The dialysis is im-

portant because it removes endogenous substrates. The enzyme preparations can be stored at 4°C for at least 3 days or kept frozen (e.g., at -80°C or at -15°C if a -80°C is not available) for a longer period of time. The same enzyme preparation can be used for Experiment 14.

Lowry assay

See Appendix D. Depending upon the amount of cells, breakage, and volume of buffer used, there may be as much as 20 to 30 mg/ml of protein.

Enzyme assay

This assay is done at room temperature. The reaction mixtures can be made by making 10x stock solutions and mixing them prior to use. Keep the enzyme solutions cold. The A-2 solution should be kept cold but allowed to come to room temperature before adding the substrate. You will also be examining the inhibition by AMP, so read the entire procedure before you begin.

1. Add 5 ml of room temperature solution A-2 to a colorimeter tube.

2. Add 0.5 mg of supernate protein (i.e., the supernatant fluid from the centrifuged cell-free extract) and mix.

3. Take readings of A_{340} every 30 sec for 2 to 3 min until the readings stablilize (or increase very slowly).

4. Start the reaction by adding 0.4 ml of 1 mM fructose-1,6-bisphosphate, and take readings every 30 sec.

The A_{340} increase is due to the reduction of $NADP^+$ by glucose-6-phosphate dehydrogenase using glucose-6-phosphate generated by the isomerase from fructose-6-phosphate (i.e., the product of the phosphatase reaction). You must obtain a linear rate in order to calculate the activity. Once a linear rate is obtained you can check the inhibition by AMP by adding AMP to a final concentration of 0.5 mM. Repeat the assay but add the AMP before adding the FBP and compare the results.

PREPARATION OF DATA

Plot A_{340} versus time and determine the velocity of the reaction from the linear portion of the curve. Express the specific activity as A_{340} per min per mg of protein in the assay.

QUESTIONS

1. What might you expect the substrate concentration versus velocity curves to be like for the phosphatase with and without AMP if the AMP were an allosteric inhibitor?

2. Was there a difference in the inhibition by AMP depending upon whether it was added before or after the FBP? Rationalize the results that you obtained.

3. Write the fructokinase and fructose-1,6-bisphosphatase reactions and sum them. Do you see why these reactions must not proceed at the same rate?

REFERENCES

Foy, J. J., and J. K. Bhattacharjee. 1977. Gluconeogenesis in *Saccharomyces cerevisiae*: Determination of fructose-1,6-bisphosphatase activity in cells grown in the presence of glycolytic carbon sources. *J. Bacteriol.* **129**:978-982.

NOTES

Experiment 13. Assaying Fructose-1,6-bisphosphatase in Saccharomyces cerevisiae

87

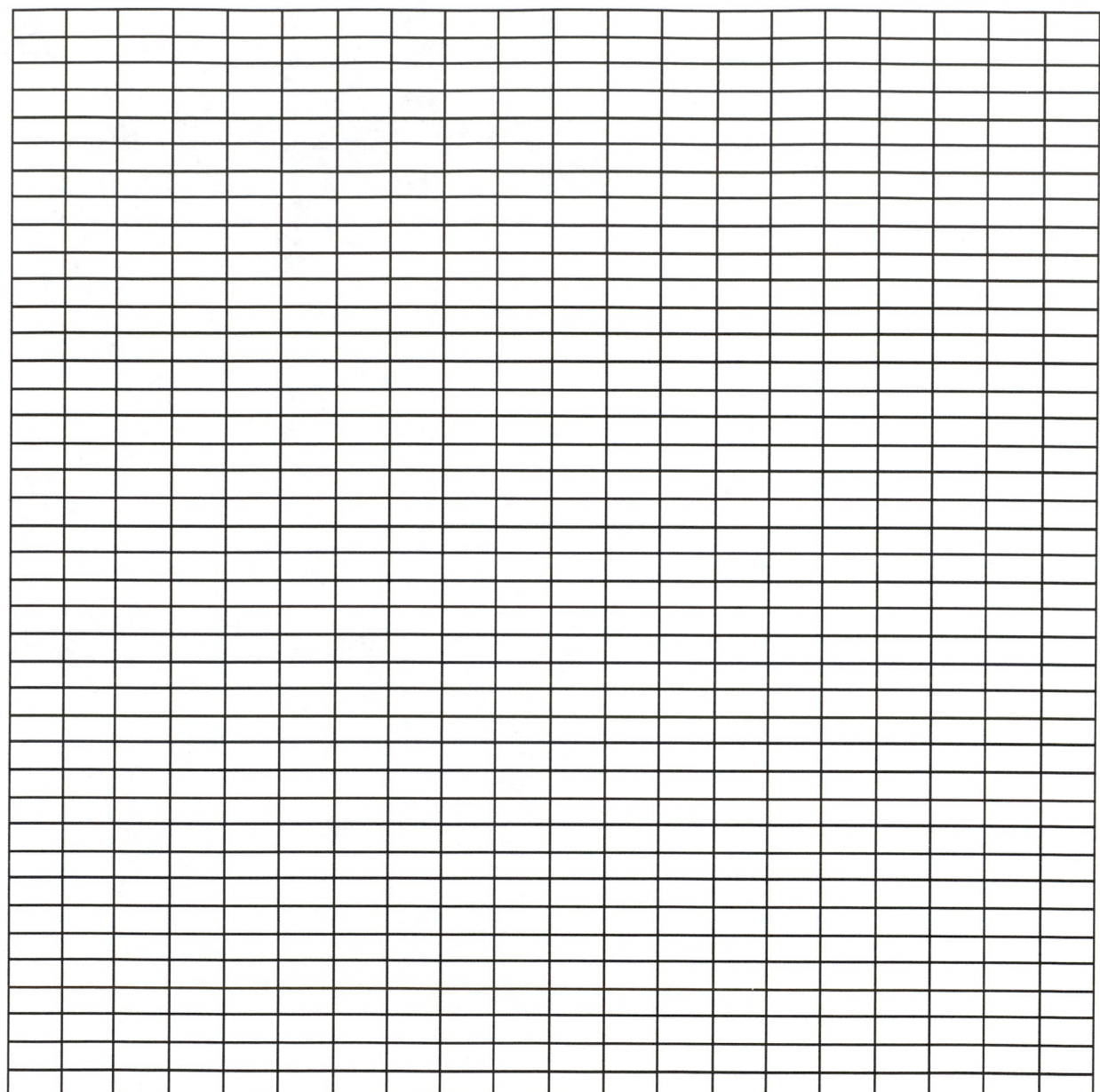

Experiment 13. Assaying Fructose-1,6-bisphosphatase in Saccharomyces cerevisiae

Experiment 14

PURIFICATION OF GLUCOSE-6-PHOSPHATE DEHYDROGENASE

Five Class Periods

Goals

The goals of this experiment are: (1) to introduce you to enzyme purification (i.e., column chromatography), (2) to teach you how to assay for glucose-6-phosphate dehydrogenase, (3) to show you how to separate polypeptides and proteins using gel electrophoresis, and (4) to show you how to stain for glucose-6-phosphate dehydrogenase after gel electrophoresis. The enzyme is stable and can be stored refrigerated in between class periods but should be frozen for prolonged storage to avoid microbial growth.

INTRODUCTION

Glucose-6-phosphate dehydrogenase catalyzes the following reaction:

glucose-6-phosphate + NADP$^+$ $\longrightarrow$

6-phosphogluconate + NADPH + H$^+$

(The 6-phosphogluconate that is produced enters either the pentose phosphate pathway or the Entner-Doudoroff pathway.) The enzyme is assayed by monitoring the increase in A_{340} due to NADPH. The enzyme will be purified by gel chromatography, and the polypeptides will be examined using gel electrophoresis.

An explanation of gel chromatography

Gel chromatography (gel filtration) separates proteins on the basis of differences in molecular size. One constructs a chromatographic bed of tiny gel beads packed into a chromatography column. The space between the gel beads is oc-cupied by liquid, which flows through the column and the proteins are carried in the liquid. The gel beads retard the migration of proteins through the column depending upon the size of the protein; the smaller proteins are retarded and the larger proteins move faster through the column. The gel bead is a cross-linked polymer. The extent to which a molecule penetrates the spaces between the cross-linked chains depends upon the hydrodynamic size of the molecule. The smallest molecules freely penetrate the spaces between the cross-linked chains and therefore are equally distributed between the liquid phase and the gel phase. The largest molecules may not penetrate the gel beads at all and are partitioned entirely within the liquid phase. These molecules emerge from the column first.

Some terms used to characterize the column

Bed. This refers to the column packing material and the interstitial fluid.

Bed height. This is the distance from the bed support to the upper gel surface.

Void volume, V_o. This is the volume of liquid in the interstitial spaces between the beads. It is determined experimentally by chromatographing a substance that is so large that it is completely excluded from the beads and measuring the elution volume. A substance that is completely excluded will elute in a sharp peak after the entire interstitial liquid is replaced. The void volume is generally measured using blue dextran, which has a molecular weight of 2 million.

Elution volume of substance, V_e. This is the volume of eluant that is required to carry the substance through the column.

The relative elution volume, V_e/V_o. This is a commonly used parameter to characterize chromatography of a substance, and it will be used in these experiments. The logarithms of the molecular weights of different proteins plotted against V_e/V_o yields a straight line and can be used to estimate the molecular weight of proteins.

Polyacrylamide gel electrophoresis

During gel electrophoresis charged molecules (including proteins) move in response to an electric field. The rate of movement depends upon a number of factors, including the strength of the electric field, the net charge on the molecule, and the size and shape of the molecule. The gels through which the proteins move are cast in tubes, slabs, or flat beds. The gel is placed between two buffer chambers, one of which contains a positive electrode (anode) and the other a negative electrode (cathode). The only electrical connection between the two chambers is through the gel. The gel particles have small pores of molecular dimensions and act as sieves. Because the pores in polyacrylamide gels are sufficiently small to restrict the flow of larger proteins relative to smaller proteins, proteins separate according to their molecular size as they move through the gel. The smaller molecules move faster.

The polyacrylamide gels are formed by mixing monomer (acrylamide) with a cross-linker, (bisacrylamide) and polymerizing them with a catalyst so that the polymerized gel consists of long chains of monomer held together by covalently linked bisacrylamide. The sum of the monomer and bisacrylamide is given as % w/v. Gels equal to 5-20% with 3-5% bis (percent of total weight of acrylamide) are generally used. The pore size decreases as the total concentration of acrylamide increases. The pore diameter also varies with the bisacrylamide concentration. Bisacrylamide at 5% gives a minimum pore size at any given concentration of total acrylamide. The polymerization is due to the addition of ammonium persulfate and N,N,N',N'-tetramethylethylenediamine (TEMED). The persulfate inititiates polymerization via free radical formation, and the TEMED acts as a catalyst.

SDS gels

Sodium dodecyl sulfate (SDS) is sometimes incorporated into the gels. The SDS denatures most complex proteins into monomers, and gives good, clean separations according to molecular weight. SDS is an anionic detergent that adsorbs to the polypeptide backbone of proteins. This denatures the proteins and gives them a negative charge proportional to their length (and molecular weight). The proteins are treated with mercaptoethanol at the same time as they are incubated with SDS. The mercaptoethanol reduces disulfide bonds that hold polypeptides together. The result of this treatment is that the proteins are transformed into SDS-coated rods with equal charge per unit length. There is a linear relationship between the log of the molecular weight of a protein and its R_f (the ratio of the distance the protein migrates from the top of the gel to the dye front or to any fast-moving reference marker).

Nondenaturing gels

This system does not use SDS and is designed to preserve biological activity and native structure of the proteins. The proteins are separated by their net charge as well as molecular weight. Separations are determined by the pH of the buffer.

Staining the gel

Protein gels are usually stained with either Coomassie blue or silver nitrate. The Coomassie blue stain detects about 1 μg of protein, whereas the silver stains are sensitive to much lower amounts. Sometimes it is possible to stain a protein by assaying for enzymatic activity.

Resistance, voltage, current, power, and heat

Ohm's Law states: $I = E/R$, where I is the current (amps), E is the voltage (volts), and R is the resistance (Ohms). The relationship between power, voltage, and current is: P(watts) = E (volts) x I(amps). During electrophoresis, one holds either current or voltage constant. If the current (I) is held constant, then the velocity of migration through the gel is constant, but heat is generated since the voltage will increase as the resistance increases during electrophoresis (heat generated = I^2R). If the voltage is held constant, then additional heat will not be generated during the run, but the current, and therefore the velocity of migration, will decrease as the resistance increases.

MATERIALS

Supplies

- S-300 Sephacryl (fractionation range 10,000 to 1.5×10^6)
- molecular weight markers for Sephacryl; example, one can use apoferritin (m.w. = 4.4×10^5), bovine serum albumin (m.w. = 6.6×10^4), cytochrome c (m.w. = 1.25×10^4), and aprotinin (m.w. = 6.5×10^3; molecular weight marker kits can also be puchased
- Blue Dextran (m.w. = 2×10^6) (Pharmacia)
- molecular weight markers for gel electrophoresis; low- and high-molecular weight marker kits can be purchased
- glucose-6-phosphate dehydrogenase from yeast
- a glass column for constructing a 200 ml Sephacryl column
- fraction collector
- spectrophotometer
- minislab gel kits

Solutions for acrylamide gels These are to be prepared by instructors. *Caution: Acrylamide is a toxin. Do not breathe the dust.*

Solution A. Monomer solution (30% T, 2.7% bis)

acrylamide	29.2 g
bis	0.8 g
water	to 100 ml

Can be stored in refrigerator in a dark bottle for about 1 month.

Solution B. 1.5 M Tris buffer, pH 8.8. (For

nondenaturing gels, omit SDS.) Dissolve 18.17 g of Tris base and 0.4 g of SDS in water. Adjust pH to 8.8 with HCl. Add water to 100 ml.

Solution C. 0.5 M Tris buffer, pH 6.8. (For nondenaturing gels omit SDS.) Dissolve 6.06 g of Tris base and 0.4 g of SDS in water. Adjust to pH 6.8 with HCl. Add water to 100 ml.

10% SDS

> SDS 50 g
>
> water to 500 ml

Store at room temperature.

10% ammonium persulfate (solution D). Add 1 ml of water to 0.1 g ammonium persulfate on day of use.

Treatment buffer 2x for denaturing gels (0.125 M Tris-Cl, pH 6.8, 4% SDS, 20% glycerol, 10% 2-mercaptoethanol)

> Tris 2.5 ml of solution C without SDS
>
> SDS 4.0 ml of 10% SDS
>
> glycerol 2.0 ml
>
> 2-mercaptoethanol 1.0 ml
>
> water 0.5 ml

Treatment buffer 2x for nondenaturing gels
Same as for denaturing gels except omit SDS and mercaptoethanol. (Use water instead.)

Divide into aliquots and freeze.

Electrophoresis buffer. 0.025 M Tris, 0.192 M glycine, 0.1% SDS. (For nondenaturing gels omit SDS.)

Tris	3.0 g
glycine	14.4 g
10% SDS	10 ml
water	to 1000 ml

Marker stain. 0.1% Bromphenol blue.
Place 1 drop into tube containing 1 ml of sample.

Protein stain (0.05% Coomassie blue R250)

isopropanol	500 ml
glacial acetic acid	200 ml
water	1300 ml
Coomassie blue	1 g

Store at room temperature.

Destain. 10% acetic acid

Stain for glucose-6-phosphate dehydrogenase

NADP+	4 mg
glucose-6-phosphate	15 mg
phenazine methosulfate	2 mg
nitro blue tetrazolium	2 mg
water	7.6 ml
solution B without SDS	0.4 ml

Make just before use. Will need about 2 ml for an 8 x 9 cm gel. *This stain is very light sensitive. It should be kept in a covered container prior to use.*

Standards

Glucose-6-phosphate dehydrogenase. Make two solutions at 0.2 mg/ml in 0.05 M potassium phosphate buffer, pH 7.4. One solution is mixed 1:1 with 2x treatment buffer without SDS or mercaptoethanol. The second solution is mixed 1:1 with 2x treatment buffer with SDS and mercaptoethanol. The sample with SDS is heated for 30 sec in a boiling water bath. Each team will use 40 μl of each standard per gel (20 μl in each of 2 wells).

Other standards. Kits of mixtures of prestained standards can be purchased, or individual standard proteins can be used. Individual standard proteins can be mixed 1:1 with treatment buffer to give a final concentration of 0.5 mg/ml. 20 µl can be applied per well. Samples with SDS must be heated for 30 sec in a boiling water bath.

Buffer for washing cells, making extracts, and running Sephacryl column

0.5 M potassium phosphate buffer, pH 7.4. This is a 10x solution. Make 200 ml.

Buffer for the enzyme assay

A-1 (See Experiment 13.)
0.1M KCl
10 mM $MgCl_2$
1 mM K_2EDTA
50 mM imidazole buffer, pH 7

Can make a 10x stock solution. On day of use, add $NADP^+$ to 0.25 mM.

Solutions and directions for Lowry and Bradford assays

See Appendix D.

Cells and extracts

See the procedures for making enzyme extracts of *Saccharomyces cereviseae* in Experiment 13. Follow those procedures to obtain a dialyzed supernate. The enzyme preparation can be stored for at least 3 days at 4°C. Prolonged storage should be in a freezer (-80°C, if convenient).

Making the Sephacryl column

Prior to Day 1 construct a 200 ml S-300 Sephacryl column. (Fractionation range, 10,000 to 1.5 x 10⁶.) This can be done during an earlier class period as a demonstration. Equilibrate the column overnight in the cold room by washing

in 0.05 M potassium phosphate buffer, pH 7.4. The next day load molecular-weight markers in 5 ml of buffer and collect 3.0 to 4.0 ml fractions. The molecular weight markers can be: bovine lung aprotinin (7.5 mg), cytochrome c (7.5 mg), bovine serum albumin (10 mg), and Blue Dextran 2000, 6 mg). The void volume is estimated with the Blue Dextran. Dissolve the Blue Dextran first in a small beaker with constant stirring. This will take around 10 min. Then filter the solution removing any particles of Blue Dextran that may not have gone into solution. (All of it should go into solution if stirred longer e.g., for 30 min.) The void volume should be approximately 60 ml. The aprotinin should elute at about 140 ml. These numbers will vary depending upon the size of the bed.

PROCEDURE - DAY 1

Absorbancy readings

Take absorbancy readings of fractions of Blue Dextran and molecular weight standards eluted from the Sephacryl column. Measure A_{280} for protein and A_{625} for Blue Dextran. (Blue Dextran also has an absorption peak around 280 nm which has about 2.5 times the absorbance at 625 nm.). Measure the cytochrome c at A_{550} as well as at 280 nm. All of the samples should be measured at 280 nm but only the blue colored solutions need be measured at 625 nm and only the orange-red solutions (cytochrome c) need be measured at 550 nm. The data will be graphed as log M.W. versus the relative elution volume (V_e/V_o) where the abscissa will have increasing values from 1.0 as V_e increases. (The void volume (V_o) is the volume required to elute the peak fraction of the Blue Dextran. Therefore, V_e/V_o

for blue dextran will be 1.0.) When estimating V_e use a point 50% up the leading edge of the peak or extrapolate the leading edge to the baseline. Using the crest of the peak is not always a good idea because the values will vary depending upon the volume of the sample applied to the column.

Loading column with cell extract

Load the column with cell extract (5-10 ml). Run overnight and collect 3 to 4 ml fractions. Make certain that you reserve at least 0.5 ml per team of the crude enzyme for protein assay, enzyme assay, and gel electrophoresis. Freeze the preparations (preferably at -80°C) if they are to be stored more than 3 or 4 days.

PROCEDURE - DAY 2

Enzyme assay

Assay the fractions for enzyme to locate the enzyme. Each team can assay a different portion of the column. Initially, every other tube can be assayed to locate the peak of enzyme. Then, if desired, all the tubes within the peak containing enzyme can be assayed. The crude fraction (i.e., before column chromatography) can also be assayed for enzyme activity at this time. After the enzyme assay, a protein assay will be done in order to determine the specific activity of the column fractions as well as the crude enzyme.

1. Add to a colorimeter tube:
   a. 3 ml of room temperature A-1 buffer with NADP+
   b. 75 µl of the crude cell extract i.e., before column, or 200 µl of the samples from the column
2. Blank in spectrophotometer with this solu-

tion. Take readings every 30 sec for 2 to 3 min to obtain a rate of A_{340} increase independent of added glucose-6-phosphate. There should be little or no increase.

3. Start the reaction by adding 175 µl of 1 mM glucose-6-phosphate. Monitor the increase in A_{340} at 15-30 sec intervals for 2 to 3 min or until the reaction rate is no longer linear. Plot the absorbance versus time in order to calculate the rate. Calculate the specific activity of the enzyme as increase in A_{340} per min per mg of protein after you do the protein assay. You can also express the activity as µmoles or nmoles of NADPH produced per min per mg protein, using a millimolar extinction coefficient for NADPH of 6.22, if the appropriate instrument is used, the light path of the cuvette is 1 cm, and the volume is 1 ml. (See "Calculating the enzyme activity".)

Protein assay

You will determine protein in two ways i.e., A_{280}, and using either the Lowry or Bradford assay. Read fractions at A_{280} and plot the data to find the protein peaks. Initially, you will have to read the absorbance of the crude material i.e., the material that was not chromatographed. You will have to first dilute it 1:100 because the absorbance is too high. You can assume that the absorbance of the first peak of protein off the column is approximately 0.25 that of the crude; that is, it has been diluted about fourfold during chromatography. There will be protein eluting after the major peak. It will have an absorbance 10-20% of the major peak. If you choose a dilution that will give readings of the major peak of 1.5 to 1.7 you will be able to read the minor peaks with the same dilution. Either a Lowry assay or a Bradford assay should be performed on every other tube to see how the protein correlates with the A_{280} readings and also to allow you to calculate the specific activity of the enzyme. A protein assay should be done on every tube that contains enzyme. See Appendix D for instructions for the Lowry and Bradford assays.

Calculating the enzyme activity

You can express the enzyme activity as A_{340}/min, which is obtained from the initial slope of the line graphed after the enzyme assay. If you divide this number by the total amount of protein in the assay tube, then you can express the activity as specific activity i.e., A_{340}/min/mg protein. If a spectrophotometer and cuvettes with the appropriate geometry are used, then you can also express the amount of NADPH formed in µmoles. Use the following expression: $A = Ecl$ where A is the absorbance, E is the mM extinction coefficient, c is the concentration in µmoles per ml, and l is the light path (1 cm) of the standard cuvette. The mM extinction coefficient for NADPH is 6.22 liter mmole^{-1} cm^{-1}. This means that a 1 mM solution of NADPH has an absorbance of 6.22. 1 mM is equal to 1 µmole per ml. Thus, dividing the A_{340}/min by 6.22 yields µmoles of NADPH per ml formed per minute. Therefore, the specific activity may also be expressed as µmoles of NADPH/min/mg protein if the appropriate instrumentation is available. (See problems #8, 9, and 10 in Appendix F.)

Pooling the peak fractions

After a class discussion the fractions containing the highest specific activity of enzyme will be combined. The pooled fractions and crude enzyme will be stored at 4°C for gel electrophoresis (or at -80°C if stored more than a few days). You should also save the peak tubes from other column fractions for SDS gel electrophoresis to satisfy yourself that the column separated the yeast proteins according to their molecular weights.

Calculating the molecular weight

The molecular weight of the enzyme will be estimated later from the relative elution volume(V_e/V_o) and data from Day 1.

PROCEDURE - DAY 3

Protein assay of pooled peaks

The pooled fractions and the crude fraction will be assayed for protein by the Lowry assay or Bradford assay. (See Appendix D for instructions.) Use 10 and 50 µl of pooled fractions and 5 and 10 ul of crude material for the protein assay. Samples can be kept at 4°C until analyzed via gel electrophesis (Day 5).

Enzyme assay of pooled peaks

Assay the unfractionated enzyme as well as the pooled peaks. Use 75 µl to assay the unfractionated enzyme and 200 µl to assay the pooled peaks.

Calculating the percent recovery

One can calculate the % recovery of enzyme activity by comparing the total activity added to the column and the total activity recovered in the peaks. The total enzyme activity in a sample is the enzyme activity per ml multiplied by the total number of ml of sample. If you assayed 0.075ml to get the enzyme activity and there are 10 ml of enzyme, then multiply the activity by 10/ 0.075 or 133 to get the total activity.

The fold purification

The fold purification is equal to the specific activity of the purified enzyme divided by the specific activity of the crude enzyme.

PROCEDURE - DAY 4

Discussion

Be prepared to discuss (1) the molecular weight of the enzyme (2) the percent recovery of protein in the pooled peaks (3) the percent recovery of total enzyme activity from the column and (4) the fold purification.

PROCEDURE - DAY 5

Overview

Both denaturing gels (SDS gels) and nondenaturing gels (no SDS) will be run on the crude extract and on material from the column. In addition, molecular-weight standards for gel electrophoresis and pure glucose-6-phosphate dehydrogenase from yeast will also be chromatographed. The denaturing gel will be stained with Coomassie blue. An activity stain will be used to locate the enzyme on the nondenaturing gel. *Caution: Unpolymerized acrylamide is a neurotoxin. Gloves should be worn at all times. Do not mouth pipette.*

Casting the gel

Students will make a 12% separation gel and a 4.5% stacking gel. The denaturing gels and the nondenaturing gels are made in the same way. The only difference between the two gels is that

there is no SDS in any of the solutions or in the electrophoresis buffer used for the nondenaturing gels. The glass sandwich and acrylamide should be placed in a pan or a baking dish in case there is leakage or spillage. Absorbent paper should be used to line the pan if available.

1. Assemble glass plate sandwich and mark the notched plate about 2 cm from the top. This will be the height to which the separation gel will fill the glass sandwich (Fig. 14.1).

2. Mix the separation gel in a small beaker (e.g., 100 ml). Once you have added solution D and TEMED you have only 5 to 10 min to complete pouring the gel. At that time the gel will begin to polymerize. For a 12% gel use 7.2 ml solution A, 4.5 ml solution B, 6.3 ml water, 70 μl solution D, and 35 μl TEMED. Add solution D and TEMED last and gently swirl. If you shake the flask, then too much oxygen may get in and the gel may not polymerize. If it is very humid, increase the TEMED by about 1/3.

3. Add separation gel solution to the mark. This can be poured from a small beaker if the glass sandwich is tipped (Fig. 14.2).

4. Overlay about 1 cm of water before the gel polymerizes. Add the water carefully using a Pasteur pipette and run the water down the side.

5. Wait 30 to 40 min for the gel to polymerize fully. Polymerization in the glass sandwich is accompanied by a sharp line that forms between the water layer and the gel. Polymerization is complete when there are no wavy patterns in the gel. The unused gel in

the beaker or flask (or in a Pasteur pipette) should polymerize in about 10 min.

6. While the separation gel is polymerizing, prepare the stacking gel but without solution D or TEMED. The stacking gel contains 0.9 ml solution A, 1.5 ml solution C, and 3.6 ml water. It is convenient to use a small beaker.

7. After the separation gel has polymerized, pour off the water overlay and drain excess water with a Kimwipe. the Kimwipe can be inserted between the glass plates to draw off the water.

8. Add 18 µl of solutionD and 15 µl of TEMED to the stacking gel solution and swirl gently.

9. Put some of the stacking gel solution on top of the separating gel and then place the comb in, being careful to avoid the formation of trapped air bubbles. Now with a Pasteur pipette add stacking gel solution until the apparatus is full (Fig. 14.3).

10. Wait about 20 to 30 min for the stacking gel to polymerize.

Placing the gel in the apparatus

1. Carefully remove the comb (Fig. 14.3). Rinse the wells with electrophoresis buffer to remove any unpolymerized acrylamide. This is an important step.

2. Remove the clips and the gasket. Be careful not to tear the gasket (Fig. 14.4).

3. Place the glass sandwhich into the gel box. Fill the well with buffer. Add electrophoresis buffer to the lower tank until the buffer is approximately 2 to 3 cm above the bottom of the gel (Fig. (14.5). There may be air bubbles at the bottom of the sandwich. These should be removed with a Pasteur pipette shaped like a J, and a rubber bulb. Forcing liquid against the air bubbles pushes them away. Depending upon the gel box, removal of the air bubbles may have to be done before the glass sandwich is secured in the rig. Secure the gel in the rig and fill the top tank with sufficient electrophoresis buffer so that buffer flows into the wells. There should be at least 1 to 2 cm of buffer above the well.

Sample preparation for SDS gels

Into a 6 x 50 mm test tube, pipette the following:

1. Equal volumes of treatment buffer and sample. The final concentraiton of protein should be 1 - 3 µg/µl (1-3 mg/ml). If the protein concentration is greater than 3 mg/ml, then use a smaller portion of the protein sample and make up the difference in volume with water.

2. Mix.

3. Place in boiling water bath and heat for 30 sec. The tubes should be covered (e.g., with a marble).

4. Add 1 drop (50 µl) of 0.1% bromphenol blue per ml of sample and mix. The bromphenol blue is a tracking dye. If the volume of sample is less than 1 ml, you can add proportionally less dye, e.g., 10 µl per 200 µl of sample. You do not want to add too much dye per volume of sample because it is important not to dilute the glycerol in the treatment buffer too much.

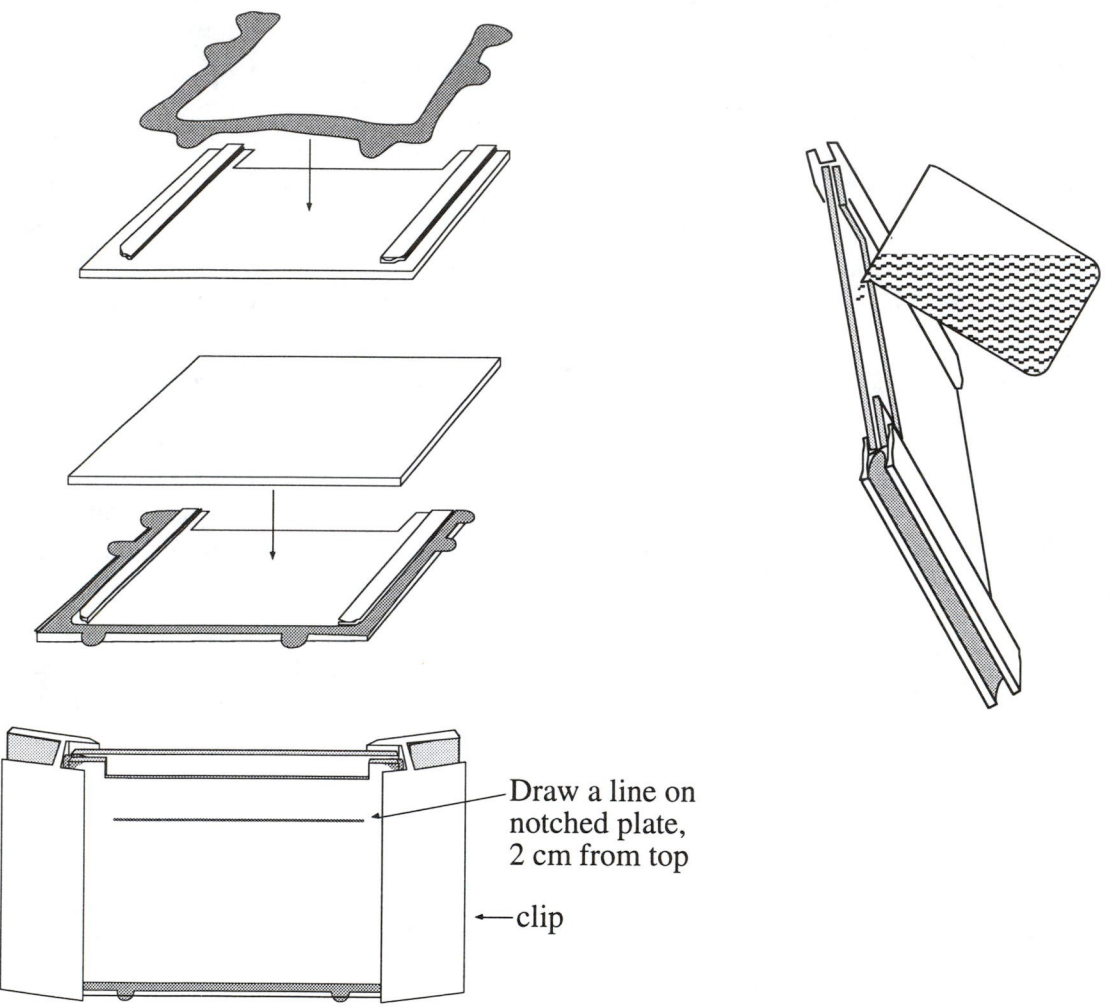

Fig. 14.1 Assembling the plates. (Adapted from operating instructions for Dual Mini Slab Kit AE-6450 from Atto Corporation, Tokyo, Japan.)

Draw a line on notched plate, 2 cm from top

← clip

Fig. 14.2 Pouring the gel. (Adapted from operating instructions for Dual Mini Slab Kit AE-6450 from Atto Corporation, Tokyo, Japan.)

Sample preparation for nonSDS gels

1. Treatment buffer does not contain SDS or mercaptoethanol.

2. Mix equal volumes of treatment buffer and sample. The final concentraiton of protein should be 1-3 mg/ml.

3. Add 1 drop of 0.1% bromphenol blue per ml of sample and mix.

Loading the gel

Load material from all the protein peaks, including material that was not chromatographed. This will allow you to see how the column fracitonated the proteins according to their molecular weights. Using a Hamilton syringe or a microliter pipetter, load each lane with about 30 µg of protein. You should be able to detect about 2 µg of protein per band. Each well will probably hold a little less than 40 µl. Also, load 5 µl of the high-molecular-weight (2.7 µg/µl) standard and 5 µl of the low-molecular-weight standard (3.5 µg/µl) in separate wells. If you are using individual proteins as standards instead of a mixture, load about 10 µg of each protein. Load standards on either side of the samples (i.e. in lanes 1 and 12). You can also load pure glucose-6-phosphate dehydrogenase (about 2 to 5 µg) (i.e, in lanes 2 and 11). Load the same samples (except the standards), including the pure glucose-6-phosphate dehydrogenase, onto the nondenaturing gel as well as the denaturing gel. Then you can look for enzyme activity in all the samples.

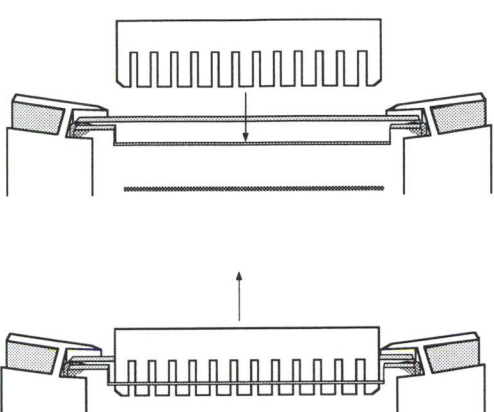

Fig. 14.3 Adding and removing the comb . (Adapted from operating instructions for Dual Mini Slab Kit AE-6450 from Atto Corporation, Tokyo, Japan.)

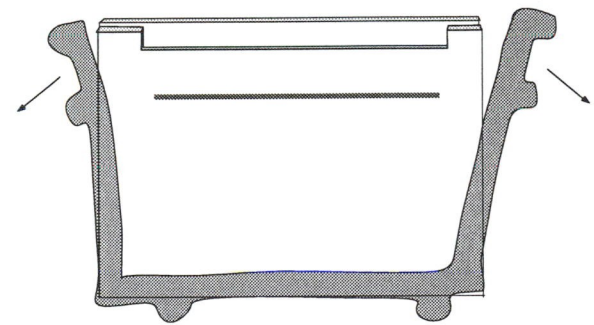

Fig. 14.4 Removing the clips and gasket. (Adapted from operating instructions for Dual Mini Slab Kit AE-6450 from Atto Corporation, Tokyo, Japan.)

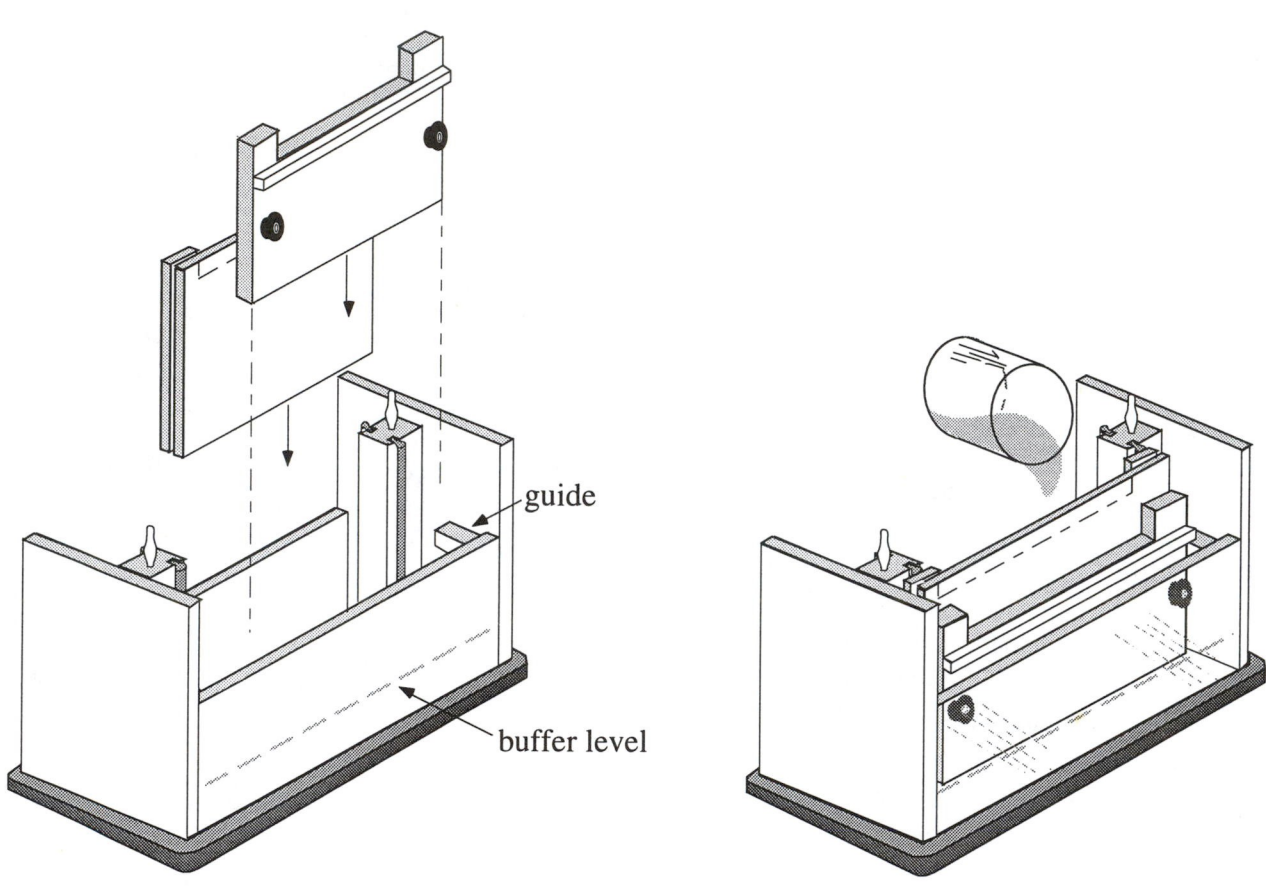

Fig. 14.5. Inserting glass sandwich into gel box and adding buffer. (Adapted from operating instructions for Dual Mini Slab Kit AE-6450 from Atto Corporation, Tokyo, Japan.)

Running the gel

Set power supply at 50 mA for the SDS gels and 20 mA for the nonSDS gels. (The lower amperage for the nonSDS gels is to ensure that the gels do not get too warm and inactivate the enzyme.) Run SDS gels about 30 min or until the dye front has traveled at least 80 to 90% of the length of the gel. If the laboratory is running late, the nondenaturing gel can be stopped when the dye front has traveled a much shorter distance.

Staining

(Wear gloves.)

1. Pour off electrophoresis buffer.

2. Pull sandwich off rig and place on bench, notched side down. Use a spatula to pry off the top plate. The gel should stick to the notched plate (Fig. 14.6).

3. Using a spatula, cut along spacers if such exist on the glass plate; otherwise the gel will stick to the spacers and will be difficult to remove from the plate. Cut a small piece of the gel on the bottom right corner in order to remember the orientation of thewells.

4. Invert the glass so that the gel faces the staining solution and carefully peel the gel off with spatula (Fig. 14.6). It should easily come off into the staining solution.

5. Stain for about 15 min at room temperature. Shake the dish periodically or place on a slow rotary shaker.

Destaining

1. Pour off stain and discard.

2. Rinse excess stain away with destain soluion. Then add 2 Kimwipes and stain to cover the gel and leave at room temperature overnight. The Kimwipes absorb the stain. The gels should be compleletely destained by the next day. They can be stored in distilled water wrapped in plastic wrap.

Staining nondenaturing gel for glucose-6-phosphate dehydrogenase

Remove the unnotched plate and pipette about 2 ml of the staining solution for glucose-6-phosphate dehydrogenase onto the gel. Incubate in the dark for 2 to 3 min or until color develops. A blue-grey region indicates formazan deposition. Rinse with 5% acetic acid to halt development. The gels can be stored in distilled water wrapped in plastic wrap.

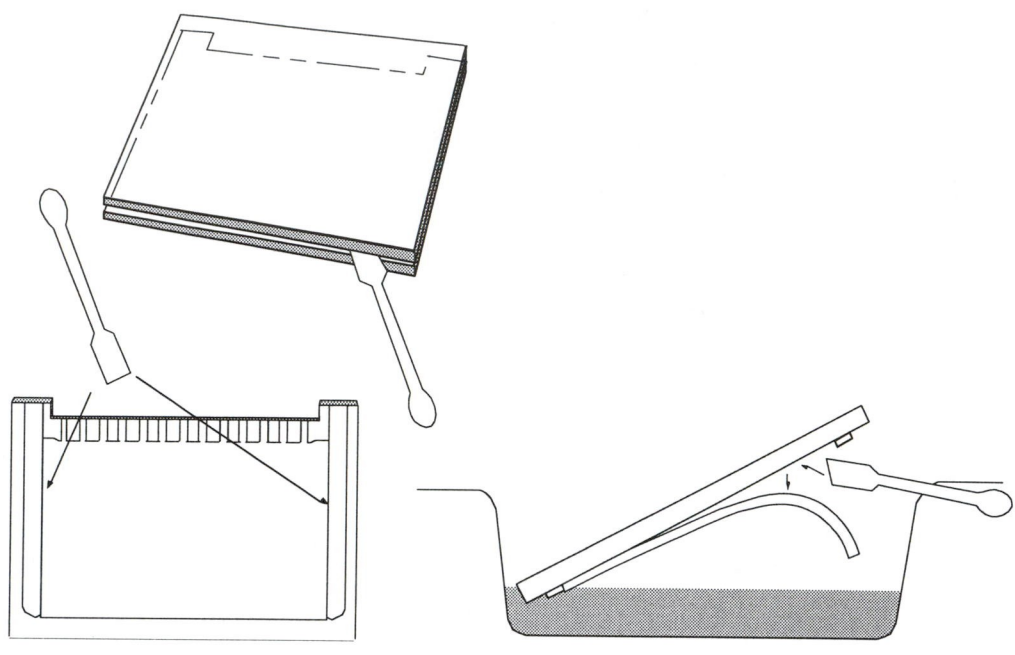

Fig. 14.6. Prying off plate and removing gel. (Adapted from operating instructions for Dual Mini Slab Kit AE-6450 from Atto Corporation, Tokyo, Japan.)

QUESTIONS

1. Calculate the fold purification and the percent recovery of activity of glucose-6-phosphate dehydrogenase after Sephacryl chromatography.

2. What is the molecular weight of the enzyme according to the Sephacryl column?

REFERENCES

An excellent description of column chromotography and gel electrophoresis, as well as other physical methods of analyses can be found in Wood, W. A. and J. R. Paterek. 1994. Physical analysis, pp. 465-511. In: *Methods for General and Molecular Bacteriology.* Gerhardt, P., Murray, R. G. E., .Wood, W. A and Krieg, N. R. (eds.). American Society for Microbiology. Washington, D.C.

NOTES

104   *Experiment 14. Purification of Glucose-6-phosphate Dehydrogenase*

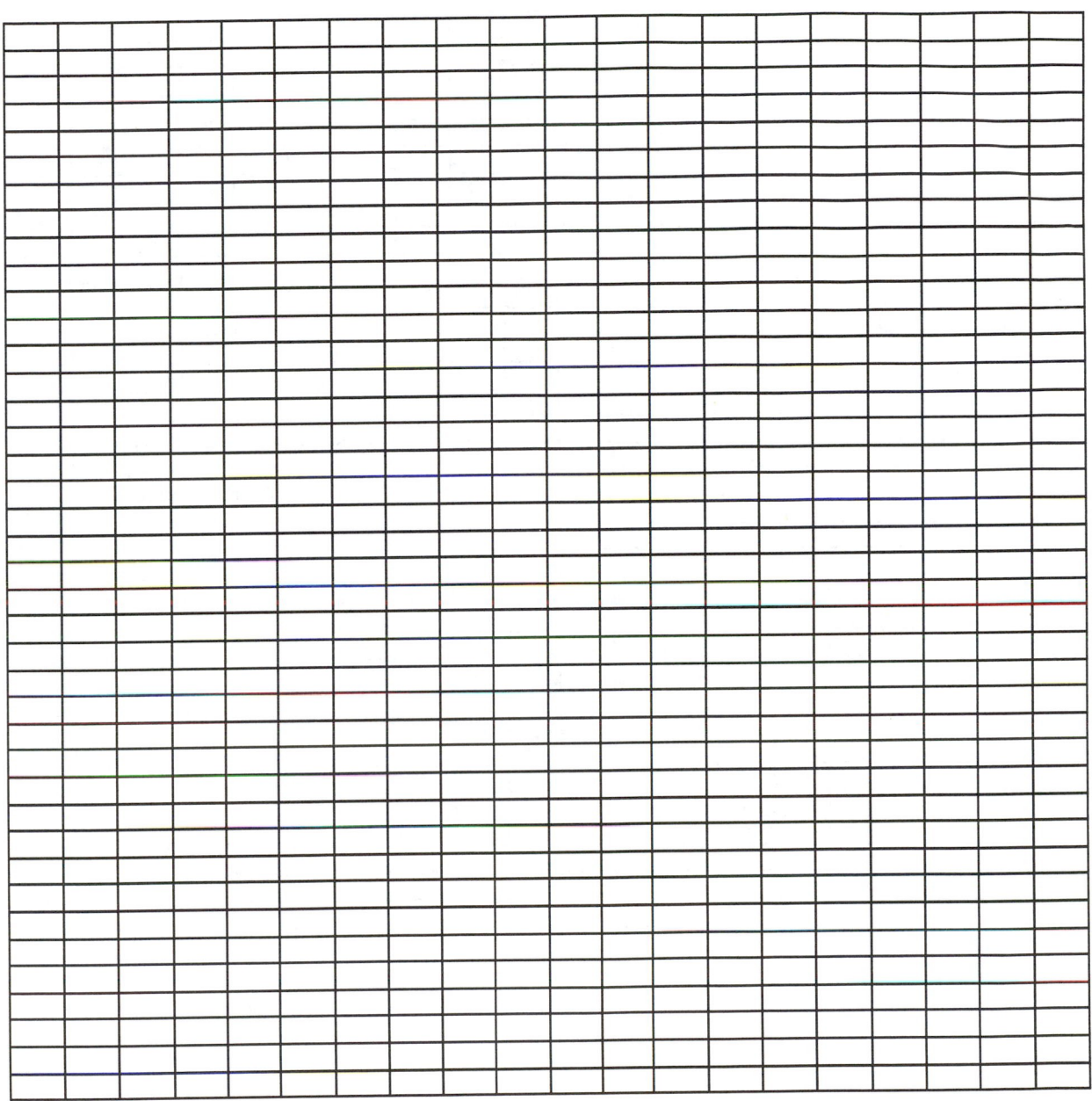

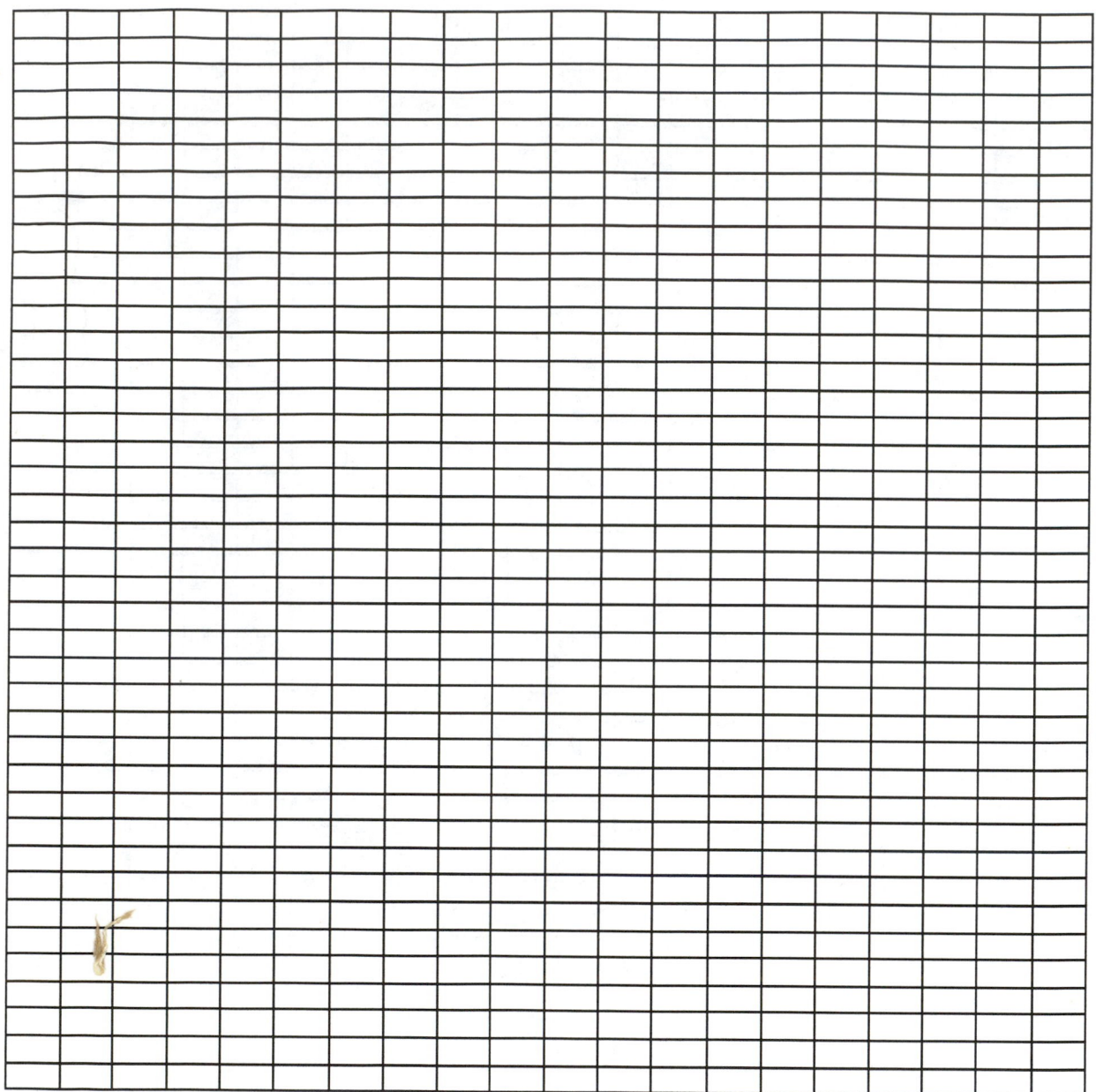

Experiment 14. Purification of Glucose-6-phosphate Dehydrogenase

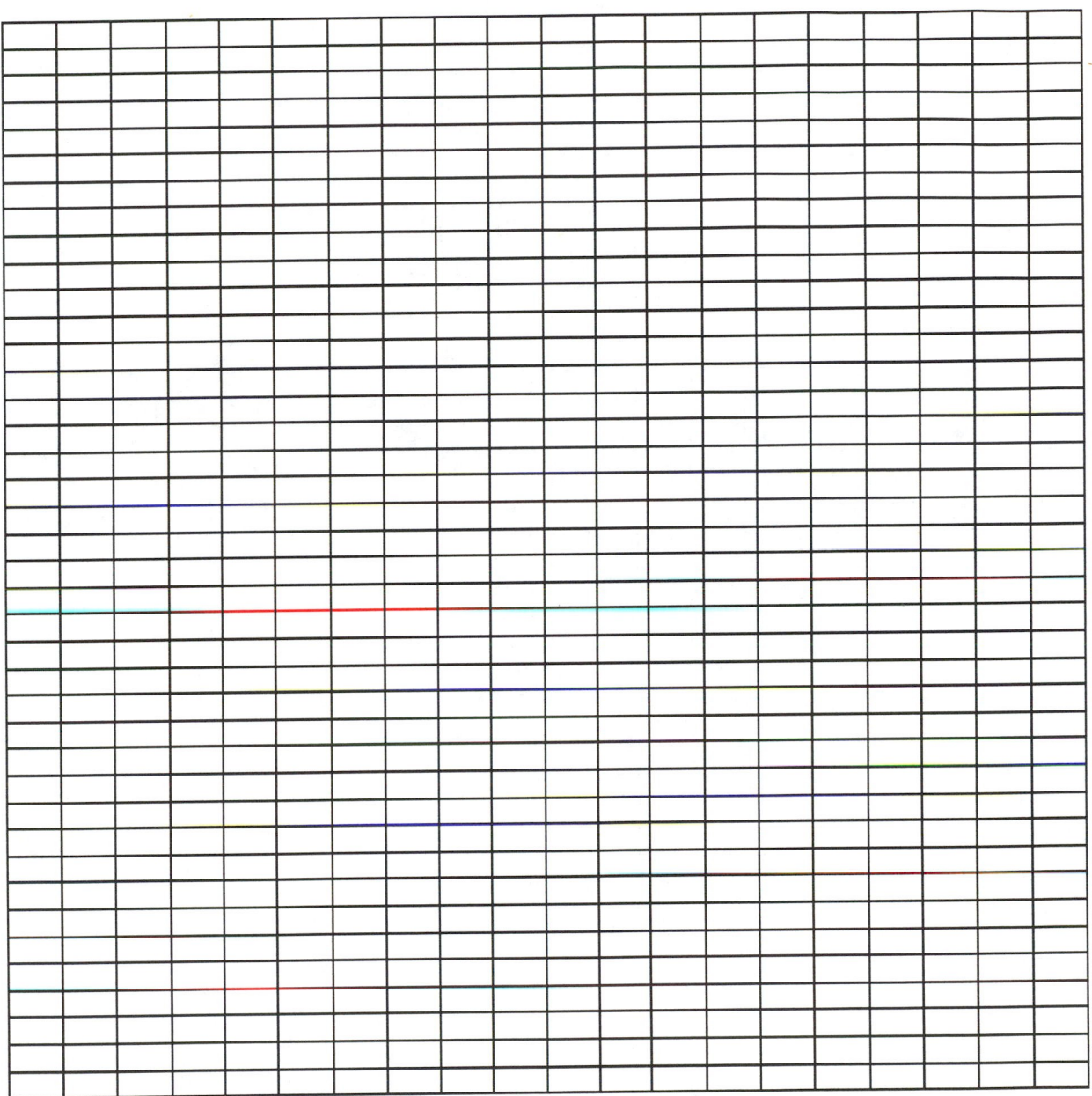

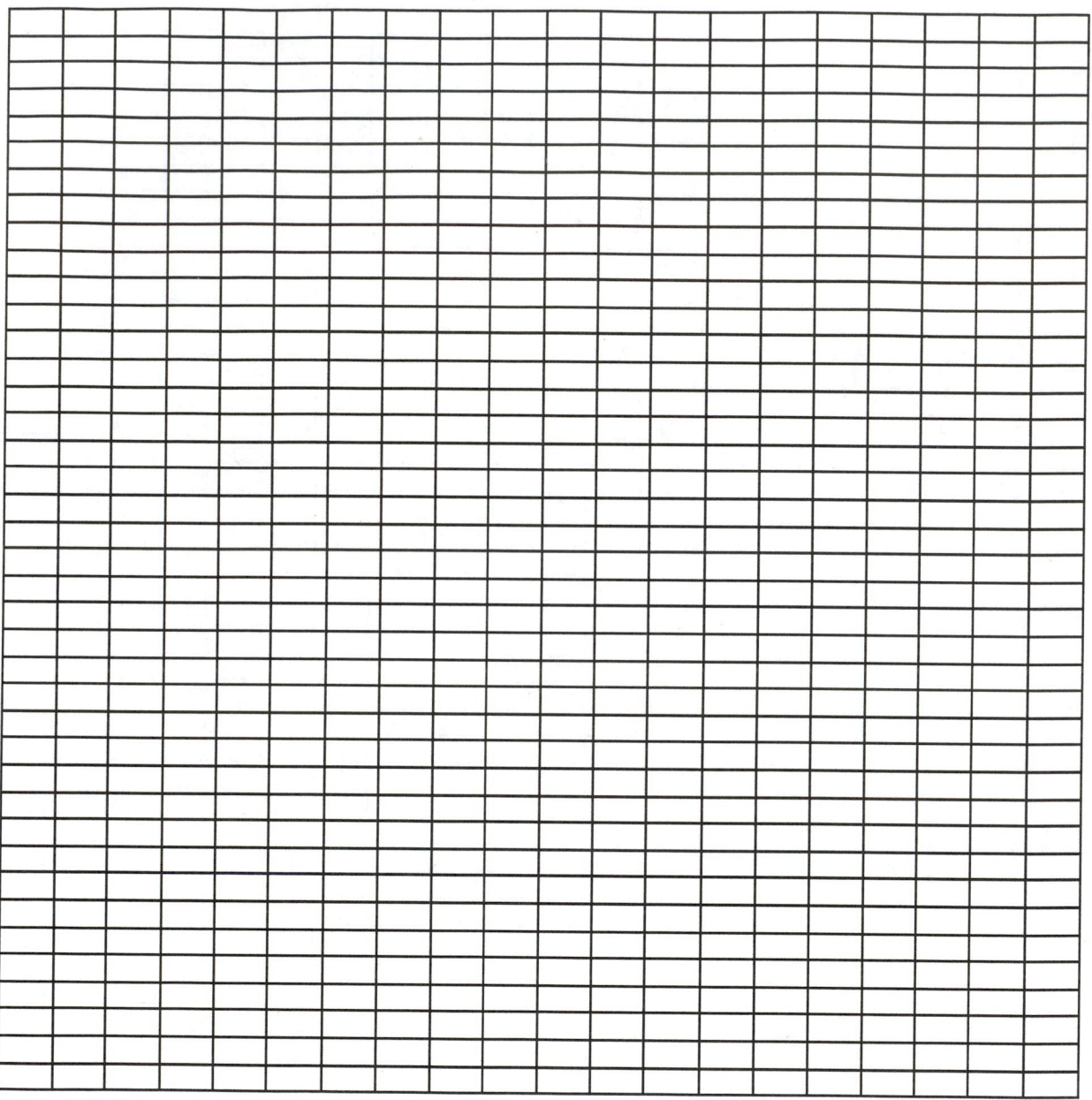

Experiment 14. Purification of Glucose-6-phosphate Dehydrogenase

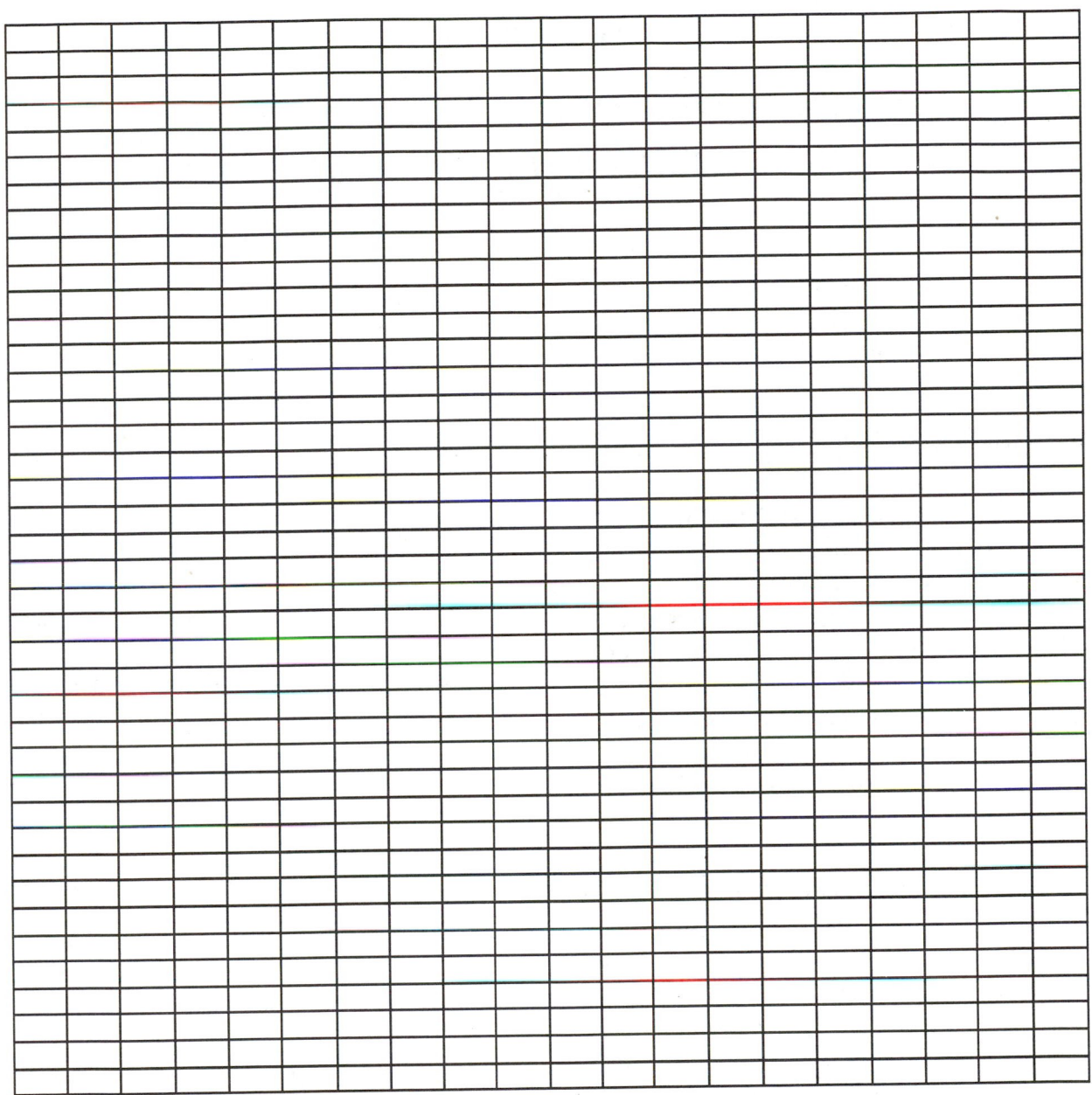

HOW *ESCHERICHIA COLI* ADAPTS TO ANAEROBIOSIS: NITRATE REDUCTASE

Goals

The goals of the experiments are to (1) learn how to assay for nitrite and (2) investigate the induction of nitrate reductase in *E. coli*.

INTRODUCTION

E. coli is a facultative anaerobe. This means that it can grow either in the presence or absence of oxygen. When growing aerobically, *E. coli* carries out an ordinary aerobic respiration sending its electrons through a cytochrome chain to cytochrome oxidase and oxygen. When growing anaerobically, *E. coli* either ferments or carries out an anaerobic respiration, depending upon the availability of carbon sources and electron acceptors. If nitrate is provided, anaerobic respiration will take place, particularly if the carbon source is nonfermentable, such as glycerol. Under anaerobic growth in the presence of nitrate many enzymological changes take place. Among these are the decreased synthesis of cytochrome *o* oxidase and the increased synthesis of respiratory nitrate reductase. These and other enzymological changes are summarized in Fig. 15.1, and the regulation of these changes by oxygen is summarized in Fig. 15.2. In this experiment, *E. coli* will be grown either aerobically or anaerobically on glucose and nitrate or anaerobically on glucose in the absence of nitrate. You will learn under what conditions the cells make nitrate reductase.

MATERIALS

Supplies and equipment

- a tank of argon for gassing out media
- 16 x 100 mm screw-cap tubes with Hungate caps (2 per student team)
- 8 gauge spinal needles for gassing out
- colorimeter set at A_{540} and colorimeter tubes
- toxic waste container

Solutions

20% sterile glucose, 100 ml.

20% (2.4 M) sterile sodium nitrate, 100 ml.

20 mM sodium nitrite, 100 ml.

Reagent A. 0.8% sulfanilic acid in 5 N acetic acid, 11 ml per team.

Reagent B. 6 ml of dimethyl-α-napthylamine in 1 liter of 5 N acetic acid *[Caution: α-naphthylamine is a carcinogen. Wear gloves when using this reagent. Discard reaction mixtures in toxic waste.]* Each team will require 7 ml for the nitrite assay.

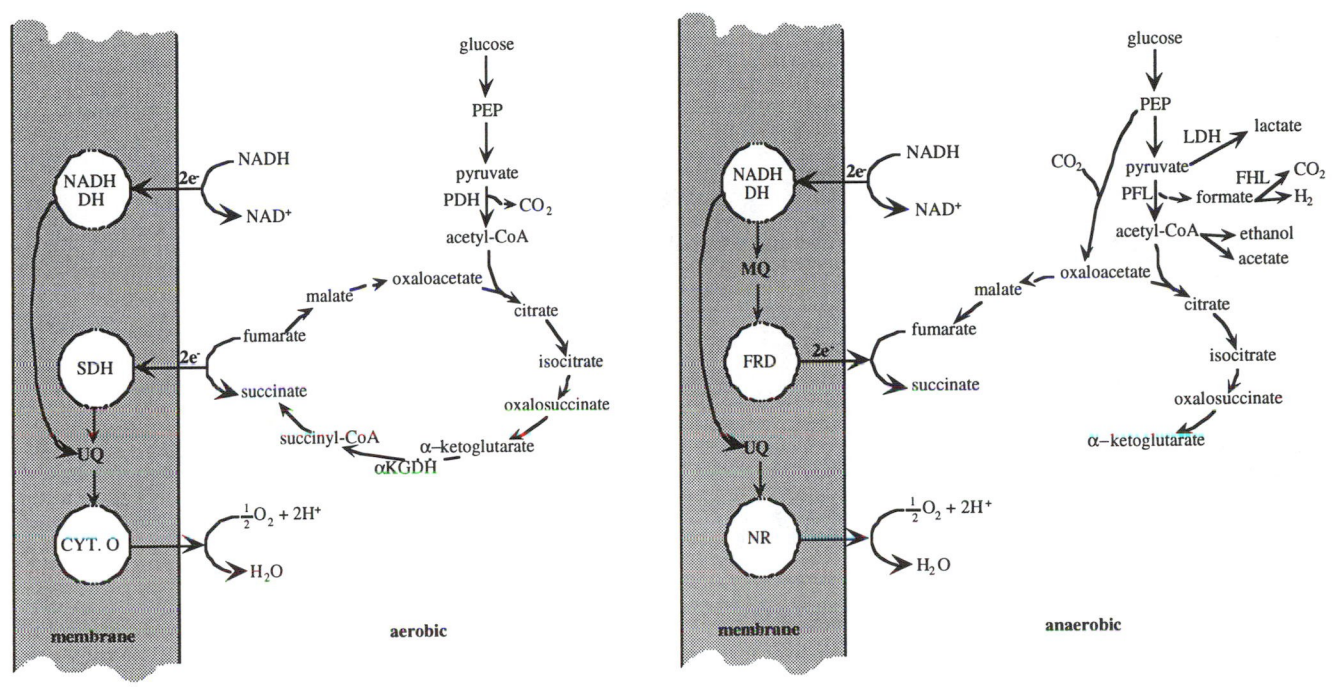

Fig. 15.1 Enzymological differences between *E. coli* grown in the presence of oxygen or nitrate. When *E. coli* is switched from aerobic growth to anaerobic growth using nitrate as the terminal electron acceptor, the following enzymatic changes occur: (a.) Cytochrome o oxidase (cyt o) is replaced by nitrate reductase (NR). (b). Succinate dehydrogenase (SDH) is replaced by fumarate reductase (FR) . Pyruvate dehydrogenase (PDH) is replaced by pyruvate-formate lyase (PFL). Alpha-ketoglutarate dehydrogenase(αKG DH) is not made. Formate-hydrogen lyase (FHL) is made. Because of the low levels of α-ketoglutaric dehydrogenase and increased levels of fumarate reductase, the citric acid cycle is converted from an oxidative cycle to a reductive pathway and succinate is made and excreted. Other metabolic changes upon anaerobic growth include the production of lactic and acetic acids, and ethanol, all of which are excreted into the medium. See Fig. 15.2 for further explanation as to how the transcription of the genes is regulated.

A

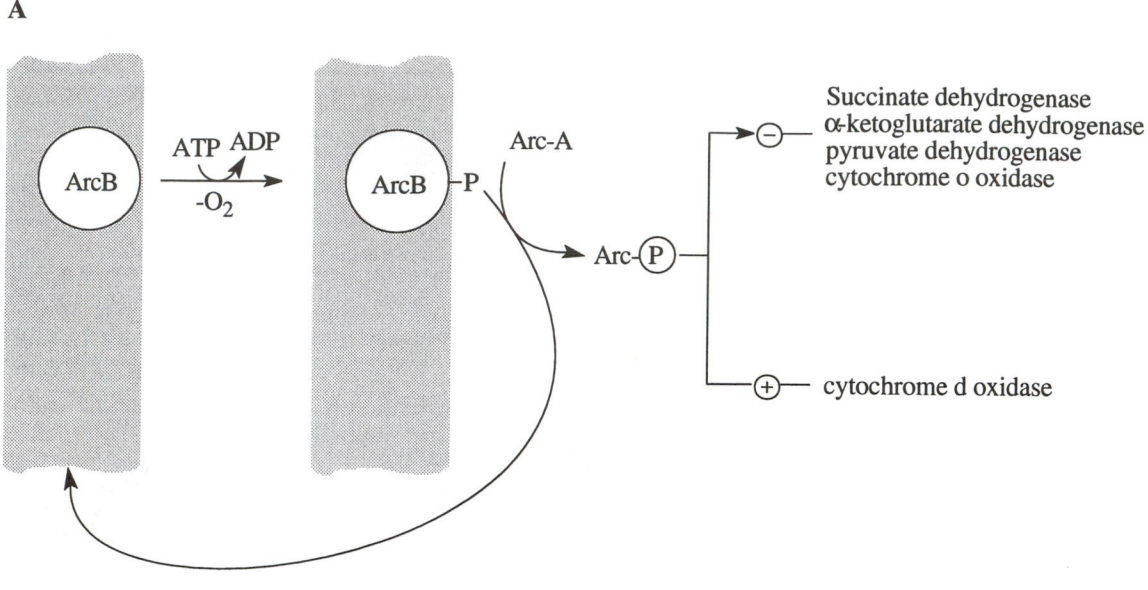

B

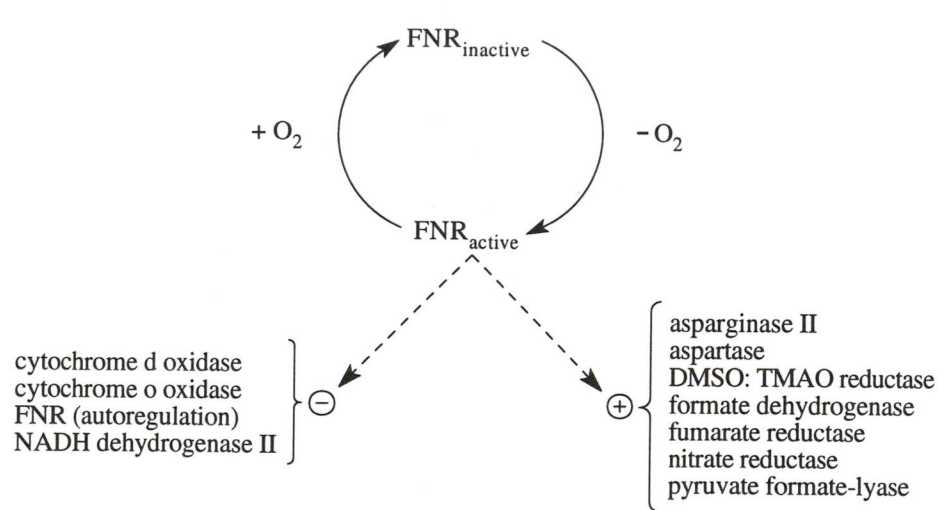

Fig. 15.2 Regulation of gene expression by oxygen and nitrate in *E. coli*. Not all genes regulated by these systems are shown. **A**. The ArcA/ArcB regulatory system. ArcB is a membrane protein that becomes activated in the absence of oxygen. Perhaps it is activated by reduction. The model proposes that the activated form of ArcB becomes phosphorylated and in turn phosphorylates a cytoplasmic regulatory protein called ArcA. The phosphorylated form of ArcA i.e. ArcA-P, is postulated to repress the transcription of genes ordinarily expressed during aerobic growth. In this way, anaerobiosis represses the transcription of many genes. Some genes are activated by ArcA-P. **B**. The FNR system. FNR is a transcription regulator that is activated during anaerobic growth. It is an inducer for many genes and a repressor for others. Together with the ArcA/Arc B system, the FNR system accounts for the differential expression of many genes in *E. coli* growth under aerobic or anaerobic conditions.

C

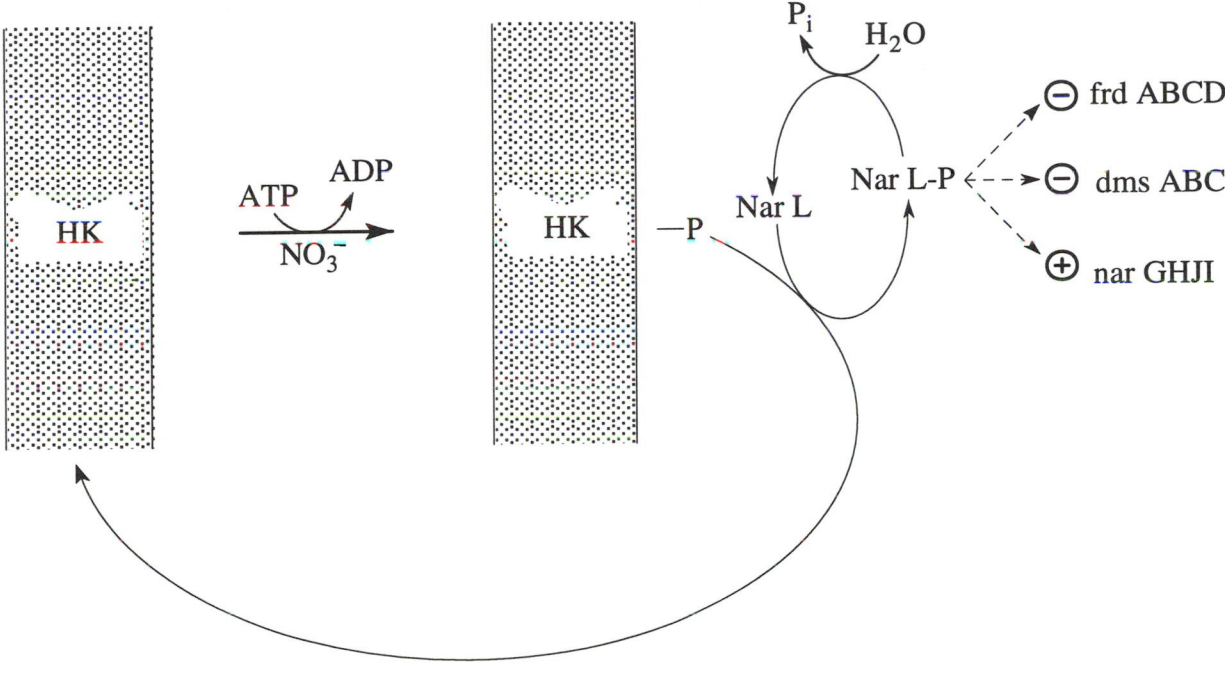

Fig. 15.2 C. Nitrate control of gene expression. A membrane-bound sensor protein called HK (histidine kinase) senses nitrate and autophosphorylates (HK-P). The phosphoryl group is then transferred to a cytoplasmic regulatory protein called NarL. The phosphorylated regulatory protein, NarL-P, stimulates the transcription of the nitrate reductase genes and represses the transcription of genes for alternative electron acceptors (the fumarate reductase and DMSO-TMAO reductase genes). (From, White, D. 1995. *The Physiology and Biochemistry of Prokaryotes*. Oxford University Press, Inc., New York.)

0.05 M phosphate buffer, pH 7.5. Each student will require 70 ml for the nitrite assay.

Media for growth of stock culture

SMB*, pH 7.0 (Appendix C) and 0.25% vitamin-free casein hydrolysate, 0.2% glucose. Use a stock 20% glucose solution that was sterilized separately and dilute it 1:100 into the medium.

Media for incubation experiments

SMB*, pH 7.0, 0.2% glucose, and 0.25% vitamin-free casein hydrolysate.

Cells

Grow *E. coli B* at 37°C shaking overnight. It is convenient to inoculate 50 ml of medium in a 250 ml Erlenmeyer flask with a loop from a plate culture.

PROCEDURE

Induction of nitrate reductase

1. Inoculate 50 ml of SMB*, pH 7.0, 0.2% glucose, and 0.25% vitamin-free casein hydrolysate in a 250 ml Erlenmeyer flask using 20µl of an overnight culture. Grow shaking (200 rpm) at 37°C.

2. When turbidity is faintly visible (around 2.5 h), transfer 10 ml to each of 2 sterile Hungate tubes.

3. Add the stock sodium nitrate to one tube but none to the other. The final concentration should be 24 mM.

4. Add the stock sodium nitrate to the flask. The final concentration should be 24 mM.

5. Replace the flask on the shaker and gas out the tube with argon for about 2 min (nitrogen can also be used. However, nitrogen gas is lighter than argon and sparging should be about 3 or 4 min.)

6. Grow samples for 1 to 2 hr until cultures are turbid (A_{660} around 0.2 to 0.3). (Growth of the anaerobic cultures uses up any oxygen that was not removed during sparging.)

Assay for NO_2^-

You will assay 3 cultures for nitrite, i.e., the aerobically grown culture and the two anaerobically grown cultures. Each assay will be done in duplicate. You will also construct a standard curve for nitrite.

1. Add 1 ml of reagent A to each of 11 test tubes.

2. Using the stock sodium nitrite, andd 0, 0.1, 0.2, 0.3, and 0.4 µmoles of sodium nitrite to 5 test tubes. This will be the standard curve.

3. Assay the 3 cultures in duplicate by adding 0.2 ml of culture to the remaining 6 tubes. (If the turbidity is too high, remove cells by centrifugation.)

4. Add 8.8 ml phosphate buffer to all the tubes.

5. Add 1.0 ml of reagent B to all the tubes.

6. Read samples at A_{540} in 30 min.

PREPARATION OF DATA

Make a table (Table 15.1) showing the amount of nitrite formed under the different growth conditions. The first column specifies whether oxygen was present, the second column whether nitrate was present, and the third column the amount of nitrite formed. Use + and - symbols to indicate whether oxygen or nitrate were present.

Table 15.1 Nitrate reductase activity

O_2	NO^-_3	NO^-_2 formed (μmoles)
+	+	
-	-	
-	+	

QUESTIONS

1. Under what growth conditions does *E. coli* make nitrate reductase?

2. What is the physiological role for nitrate reductase?

REFERENCES

Dobrogosz, W. J. 1981. Enzymatic activity, pp. 365-392. In: *Manual of Methods for General Bacteriology*. Gerhardt, P.. Murray, R. G. E., Costilow, R. N., Nester, E. W., Wood, W. A., Krieg, N. R., and Phillips, G. B. (eds.). American Society for Microbiology, Washington, D.C.

NOTES

Experiment 15. How Escherichia coli Adapts to Anaerobiosis: Nitrate Reductase

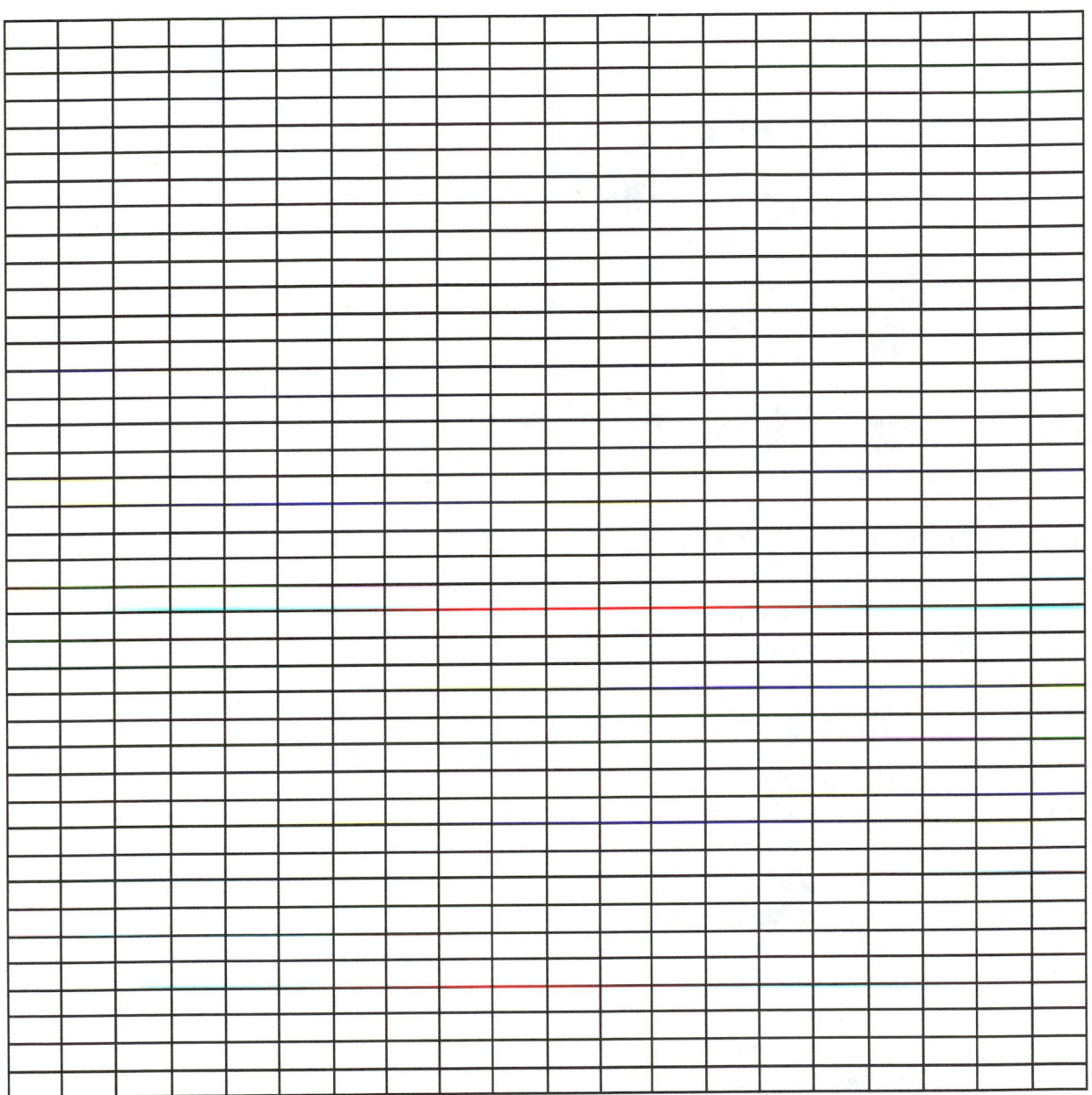

CELLULAR FATTY ACIDS OF
ESCHERICHIA COLI

Two Class Periods

Goals

The goal of the experiment is to learn how to analyze fatty acids using gas chromatography. This experiment can also be used to show how the relative amounts of saturated, unsaturated, and cyclopropane fatty acids change with growth conditions.

INTRODUCTION

Bacterial membrane lipids are primarily phospholipids consisting of fatty acids esterified to glycerol phosphate. (Archaeal membrane lipids consist of alcohols in ether linkage to glycerol or glycerol phosphate.) This experiment describes the extraction and saponification (alkaline hydrolysis to liberate the fatty acids) of the cellular lipids of *Escherichia coli B*, the preparation of the methyl esters of the fatty acids, and the analysis of the methyl esters by gas liquid chromatography (GLC).

The qualitative cellular fatty acid composition of many bacteria is known and is the basis for at least one commercial system for bacterial taxonomy. Bacterial fatty acid composition varies quantitatively with temperature of growth and as a function of, for example, alcohol (ethanol) concentration in the growth medium. In this experiment a mineral growth medium must be used since bacteria can assimilate fatty acids from complex media with consequent distortion of the cellular fatty acid composition. The student is referred to Appendix H for additional experiments that may be done to investigate fatty acid composition in *E. coli* grown under different growth conditions.

Gas liquid chromatography

Gas liquid chromatography accomplishes the separation of solutes that partition differently between a mobile gas phase and a nonmoving thin layer of liquid attached to a solid support. During gas liquid chromatography the sample is injected into a heating block. Immediately, the sample is vaporized and moved into the column inlet as a plug of vapor by the stream of inert gas (helium or nitrogen). The methyl esters adsorb to the stationary phase in the column and are desorbed by fresh carrier gas. The adsorption and desorption are repeated many times as the methyl esters move through the column and become separated from one another. A band forms for each methyl ester because they travel through the column at individual rates determined by their partitioning between the stationary phase and the inert gas. The bands exit the column and pass through a detector that is attached to a strip-chart recorder. The methyl esters are identified by the time at which the band passes through the detector (retention time or RT). The retention time depends upon the particular column used, the operating temperature, and the carrier gas flow rate. The amounts are proportional to the area under the recorded peaks.

Columns

The most widely used column for examining fatty acids is the fused silica capillary column containing as the stationary phase a cross-linked methylsilicone. This is a relatively small column (25-50 m long with an internal diameter of 0.2 to 0.4 mm) and gives excellent separation.

Detectors

Detectors that can be used include the thermal conductivity detector (TCD) and the flame ionization detector (FID). The TCD consists of four filaments, each one placed in a separate cavity in a brass block. Two of the filaments are surrounded by the carrier gas, and two of the filaments are surrounded by the column effluent. Gas flows around the filaments and through the cavities. Heat is lost via conduction through the carrier gas. The sample components in the effluent gas change the thermal conductivity of the gas, and this results in a change in temperature of the filaments. The change in temperature of the filaments results in a change in the electrical resistance of the filaments. This results in an electrical output. In the flame ionization detector there is a hydrogen flame that burns compounds entering the detector to ions. Surrounding the flame there is an electrode which is kept at a potential of at least -150 V in order to collect the ions. An electric current is produced that is proportional to the rate at which the sample enters the detector. An integrating recorder produces a trace of the peaks.

MATERIALS

Supplies

- Pasteur pipettes (9") and rubber bulbs to fit (around 5 per sample)
- 15 x 125 mm screw capped culture tubes fitted with Teflon-lined caps (4 per sample)
- 50 ml Erlenmeyer flasks and cork stoppers to fit (1 per sample)
- pH indicator paper
- 3 ml (or other small) conical centrifuge tubes with cork stoppers (1 per sample)

Equipment

- fume hood
- dry, clean gas supply (e.g., nitrogen)
- centrifuge to take 15 x 125 mm capped tubes (ca. 3000 rcf x g)
- flame ionization detector-equipped gas chromatograph fitted with appropriate column and integrator or recorder
- Hamilton syringes for sample injection (10 µl) and additions of solvents (50-100 µl).
- water bath at 65°C or a heating block
- tank of argon or nitrogen plus fittings for gassing multiple tubes. Each team will have to evaporate a sample under a stream of inert gas.

Solutions

PBS. See Appendix C (5 ml per sample).

Methanolic KOH (1:1:1 solution of 50% KOH, methanol and water) (5 ml per sample).

Ethyl ether (about 25 ml per sample).

Petroleum ether (30-60º) (6 ml per sample).

Concentrated HCl.

MgSO₄ (anhydrous).

BF₃-methanol reagent (Pierce Chemical Co., Cat. No. 49370).

Chloroform.

Saturated aqueous NaCl (2 ml per sample).

Cells

Individual 10-20 ml cultures of *E. coli B* in late log or early stationary phase grown in SMB* medium containing 0.4% glucose (10-40 mg wet weight cells). (Other mineral media such as M-9 may be used. See Appendix C.)

PROCEDURE

Sediment 15 ml of cells in a 15 x 125 mm centrifuge tube by centrifugation at 1500 x g for 10 min and pour off the supernate. Wash once by resuspension in 5 ml of PBS (vortex) and resedimentation. Discard supernatant fluid.

Saponification

(DO IN FUME HOOD WITH EYE PROTECTION)
To the sedimented cells add 5 ml of the methanolic KOH saponification reagent, mix, flush with inert gas (N₂ or Ar), cap, and incubate in a water bath at 65ºC for 1 h. Prepare a reagent blank without cells. After cooling, extract the sample with 2-5 ml portions of petroleum ether to remove the nonsaponifiable lipids. Carefully

acidify the samples with concentrated HCl (about 1.7 ml) to approximately pH 2 (indicator paper). Extract the free fatty acids with 3-5 ml portions of anhydrous diethyl ether and place these in a 50 ml Erlenmeyer flask to which about a teaspoon of anhydrous MgSO₄ has been added to remove water contained in the ether samples. Cap with cork stopper and allow to dry for 10 to 15 min with occasional stirring.

Methylation[1]

(DO IN FUME HOOD WITH EYE PROTECTION)
Transfer dried ether containing the fatty acids to a clean 15 x 125 ml tube. Use a gentle stream of gas (and a warm, not hot, water bath) to reduce the volume of ether to near dryness. Add 1.0 ml BF₃-methanol reagent and transfer to the 65ºC water bath. When it is clear that all the ether has evaporated (there should be no bubbles of ether when the tube is shaken), flush with inert gas, cap, and incubate for 10 min. Cool, add 3 ml petroleum ether and 2 ml saturated NaCl, and mix. Remove and save the petroleum ether layer in another screw-cap tube, dry with MgSO₄, and transfer to the 3 ml conical centrifuge tube. Take the sample to dryness carefully under a gentle stream of gas. Use 25-50 µl of CHCl₃ to wash down the walls of the tube. Stopper and analyze (gas chromatograph, GC) or store at -20ºC for later analysis. The samples can be stored dry and the chloroform added when the sample is chromatographed.

Analysis

(Teams should sign up for use of the gas chromatograph.) Set the temperature of the FID to 300ºC, the injector temperature to 250ºC, and program the instrument so that the column temperature increases from 170 to 300ºC at 5ºC per min, and then is maintained at 300ºC for 1 min. Remove solvent by use of a stream of inert gas and add a known amount (e.g., 50 µl) of $CHCl_3$ to the sample tube. Inject an appropriate amount (say, 2.5-5 µl) of fatty acid methyl ester (FAME) sample into the GC septum to initiate analysis. By use of appropriate commercial standards (e.g., Applied Sciences L208 and L205) the identities of the individual FAMEs may be inferred from retention times or by mixture with authentic FAMEs and co-injection. A James plot (log retention time vs. chain length) may be used to infer identity if only a few members of the saturated fatty acid series are available as methyl esters for use as standards. For FAMEs the flame ionization detector response approximates well the relative weight of the individual acids in the mixture.

Note: Bromination will remove unsaturated and cyclopropane fatty acid methyl esters from the chromatograms by reducing their volatility. In order to brominate, add a drop of Br_2 to a $CHCl_3$ solution of a part of the sample, allow to react, take to dryness with N_2, take up the FAMEs in an appropriate amount of $CHCl_3$, and reinject.

Control

An important control is missing from this experiment as so far described. It is a blank sample comprising an empty tube which should be introduced at this point and treated exactly as if it contained a cell sample. Why? Consider the contaminants that may occur in solvents, reagents, etc. How will you be able to tell which or how much of these incidental but indigenous substances affect the final chromatogram on which your analysis depends? An injection of a "solvent blank" or just a "blank sample" control will tell you immediately which peaks derive from the bacterial sample (presumed FAMEs) and which are just background contamination. Subtract the background peaks discovered in this way.

PREPARATION OF DATA

Since the response of a flame ionization detector is proportional to mass for a structurally similar class of compounds such as long-chain fatty acid methyl esters, the output of a recording integrator or a computer-based integrator program can be used directly to calculate the relative amounts (as mass) of the FAMEs directly from the chromatogram. If the recording integrator is capable of it, this is perhaps most conveniently rendered for tabulation by preparing a "% of total" for each FAME by use of the built in algorithm (most recording integrators can do this). Alternatively, the integrated detector response for each individual FAME can be divided by the total response (adding all peaks together) and multiplied by 100 to produce such a table. Any significant background revealed by the solvent blank should be subtracted from each chromatogram. This should be done before or after the above calculation is performed unless the blank is deemed negligible (area equal to or less than 2% of total area of a typical FAME sample). A separate calculation can be done to give the S/US ratio by simply adding the areas integrated for all saturated FAMEs and dividing by the sum of the areas for the unsaturated species for each sample chromatographed. For additional ease of visualization the S/US (or US/S) ratio could be plotted versus the growth temperature or alcohol content of the culture from which samples were derived. The cyclopropane acid areas (as % of total) could similarly be plotted to give the trend by which they are formed as a function of culture age or by superimposing the % of total values on the growth curve of the culture at each sample time for maximum clarity. (See Independent Projects in Appendix H.)

QUESTIONS

1. Are the results you obtained quantitative? How would you best represent them in a table (e.g., micrograms of each fatty acid or % of total?). If they are not quantitatively reliable, how could the procedure be changed to make results obtained on different days, on different samples, etc., more comparable? How could they be made truly quantitative?

ENDNOTES

1. The use of diazomethane as a methylating agent is cleaner, more convenient and in many ways more suited to use with multiple samples. Diazomethane (from, e.g., Diazald, Aldrich Chemical Co.; apparatus especially for preparation of this reagent is also available from this source) in ether is simply added in a hood to the dried ethereal samples until bubbling stops and a yellow color persists. Reduce volume of the FAME solution in ether that results under a stream of N_2 and proceed as indicated. However, diazomethane is explosive and allergenic and was considered by the authors unsuitable for class use.

REFERENCES

Descriptions of GLC and of fatty acid analyses can be found in:

James, A. T. 1960. Qualitative and quantitative determination of the fatty acids by gas-liquid chromatography, Vol. 8, pp. 1-59. In: *Methods of Biochemical Analysis*. Glick, D. (ed.). Interscience Publishers, Inc., New York.

Smibert, R. M. and Krieg, N. R 1994. Physical analysis. pp. 607-654. In: *Methods for General and Molecular Biology*. Gerhardt, P., Murray, R. G. E., Wood, W. A., and Krieg, N. R. (eds.). American Society for Microbiology. Washington, D.C.

Wood, W. A. and J. R. Paterek. 1994. Physical analysis. pp. 465-511. In: *Methods for General and Molecular Biology*. Gerhardt, P., Murray, R. G. E., Wood, W. A., and Krieg, N. R. (eds.). American Society for Microbiology. Washington, D.C.

NOTES

Experiment 16. Cellular Fatty Acids of Escherichia coli

123

Experiment 17

CHEMOTAXIS OF *PSEUDOMONAS AERUGINOSA*

One Class Period

Goals

The goals of the experiment are (1) to become familiar with the phenomenon of chemotaxis and how to measure it quantitatively, and (2) to investigate which compounds might be chemotactic for *P. aeruginosa*.

INTRODUCTION

Bacteria exhibit taxis responses toward or away from a variety of chemicals and light. Most compounds toward which bacteria move are nutrients (carbon and/or energy sources or growth factors). Significant progress has been made in elucidating the molecular basis for these tactic responses, mostly by experiments with *Escherichia coli* and *Salmonella typhimurium*. Taxis responses can be measured by photomicrographic techniques (measuring the proportion of time spent in "straight runs" versus "tumbling") or by counting the number of cells that enter a small capillary tube containing the attractant or repellent compound in a given time compared to appropriate controls. In this laboratory the latter method will be used.

Supplies and equipment

- sterile 1 µl capillary tubes (commercially available for applying samples to thin-layer chromatography plates. For example, Drummond microcaps.)
- sterile Pasteur pipettes
- 1000 µl Eppendorf pipettes and sterile blue tips
- 1 sterile plastic capillary chamber per team (fig. 17.1); each chamber will accommodate 4 tests; the chambers should be sterilized by UV irradiation and stored in sterile Petri plates
- slide warmers at 37⁰C and thermometers for the slide warmers
- spreaders and turntables for viable cell counts
- Tuberculin disposable syringes (1 ml), and Tygon tubing to fit capillary tubes to eject cells from the capillaries for doing the viable counts (fig. 17.2). Each team should have one.
- colony counters
- A small beaker of CTX buffer for rinsing the capillaries; one per team.
- For handling the capillaries: Capillaries are conveniently handled by use of No. 4 forceps over the tips of which are placed short (5 mm) lengths of Tygon microbore tubing (0.020 x 0.060 in). these "padded" forceps should not break capillaries nor allow them to twist while fitting them with tubing to collect the cells or during washing prior to cell collection. Each team should have a pair of forceps.

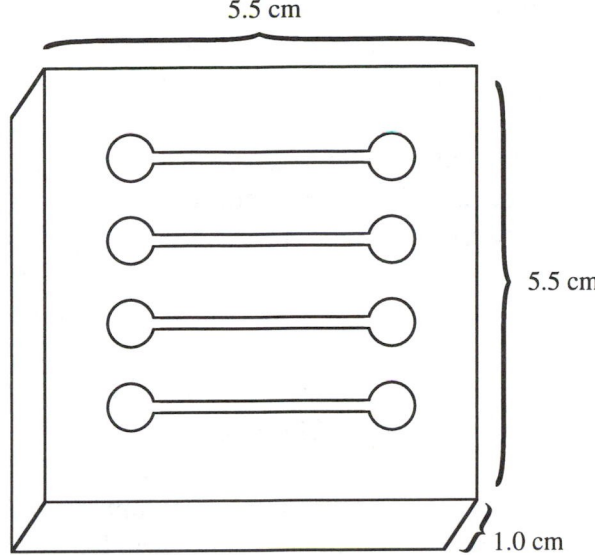

Fig. 17.1 Chemotaxis chamber. Four chambers are cut into a lucite block. Each chamber has two cylindrical compartments 7 mm in diameter and 5 mm in height, connected by a channel that is 24 mm long, 2 mm wide, and 2 mm deep. The chambers are placed in sterile Petri dishes and sterilized by UV light. (Kelley-Wintenberg, K., and T. C. Montie. 1994. Chemotaxis to oligopeptides by *Pseudomonas aeruginosa. Appl. and Env. Microbiol.* **60**:363-357.)

Solutions

Chemotaxis (CTX) buffer:
50 mM potassium phosphate [pH 7.0]
5 mM $MgCl_2$
15 µM disodium EDTA
Prepare approximately 50 ml per team and approximately 200 ml for washing the cells.

Test solutions, 10 mM each in CTX buffer
Solutions of individual amino acids, dipeptides, tripeptides, and sugars in CTX buffer will be tested. The solutions should include methionine, aspartate, valine, arginine, and serine. The solutions may be prepared by the instructor or the students.

Phosphate-buffered saline (PBS). See Appendix C. Have these ready in sterilized test tubes containing 4.9 ml and 9.9 ml as dilution blanks. Each team will require about eight 9.9 ml blanks and four 4.9 ml blanks.

Plates for viable counts. Sixteen plates of Nutrient Agar per team. (This will allow platings to be done in duplicate.)

Cells

P. aeruginosa, Preceptrol strain ATCC 10145 is grown overnight with aeration at 37ºC to the late log phase or stationary phase of growth in mineral succinate medium with ammonium ions as the nitrogen source (SMB* pH 6.8 containing 0.4% sodium succinate). Prior to class dilute the cells 1:1 with fresh medium and grow for 60 to 90 min. This insures a highly motile population. After chilling the cells on ice for 15 min, harvest by centrifugation, wash twice, and resuspend in CTX buffer at a density of about 2×10^8 cell per ml. This is an absorbance at 590 nm of 0.2. Each team will need 3 to 4 ml of cells. *CAUTION: P. aeruginosa is an opportunistic pathogen. Avoid exposure to cultures, especially if you have cuts or burns. Discard plates in the biohazard bags provided for autoclaving or, if glass, in the discard trays lid up for autoclaving prior to washing.*

PROCEDURE

Each student may make up their own solutions of attractants or repellents in CTX buffer. It is not necessary to use aseptic techniques making the 10 mM solutions provided the solutions are made just before use (why?). Test both the controls and test samples in duplicate.

Fig. 17.2 Syringe for ejecting cells from capillaries. Attach a 21 guage needle to a 1 ml disposable Tuberculin syringe. The needle is fitted with a 0.02" by 0.06" Tygon tubing about 1" long. A 1 μl capillary tube (e.g., Drummond microcap) containing the cell suspension is inserted into the Tygon tubing. If the capillary tube is too thick, then an 18 guage needle can be used to which a 0.030" by 0.09" Tygon tubing is attached.

1. Place chamber on a slide warmer at 37⁰C and add 150 μl of cell suspension to all of the wells. The channel will fill with fluid.

2. Using the forceps, fill capillary tubes with the test solution and carefully place each centrally in one of the four channels filled with the bacterial suspension. Be careful not to trap air bubbles at the opening to the capillaries as you insert the capillary centrally in the fluid-filled channel. Make certain that you include a control capillary for each chamber containing only CTX buffer. Note the number of your chamber.

3. incubate 40 min and then remove the capillaries with tweezers and rinse the outside of the capillaries with CTX buffer. This is conveniently done by dipping the capillary in the buffer. This is an imporant step (why?).

4. Expel contents of capillary into a phosphate-buffered saline (PBS) dilution blank using a Tuberculin syringe (1 ml) filled with CTX buffer and fitted with flexible plastic tubing (Fig. 17.1). Make the appropriate dilutions for viable counts. Examine Table 17.1 which will give you some guidelines for the dilutions that you should make.

5. Plate 0.1 ml of the approriate dilution in duplicate using a spreader and turntable.

6. Incubate at 37⁰C for 24-36 h . Colonies of *P. aeruginosa* tend to spread a bit; so do not wait more than 48 h to count them or to refrigerate the plates.

QUESTIONS

1. After examining the class data, can you make a generalization as to what kinds of compounds serve as attractants and what kinds serve as repellents?

PREPARATION OF DATA

Table 17.1 includes data adapted from similar experiments reported in the literature. Prepare a table similar to Table 17.1.

REFERENCES

Kelly-Wintenberg, K. and T. C. Montie. 1994. Chemotaxis to oligopeptides by *Pseudomonas aeruginosa*. *Appl. and Env. Microbiol.* 60:363-367.

Experiment 17. Chemotaxis of Pseudomonas aeruginosa

Table 17.1 Responses of *P. aeruginosa* PAO to different peptide chemoattractants[a]

Attractant[b]	CFU/capillary tube[c]	RTR[d]	RTR with respect to Arg[e]
Ala-Gly-Ala	1×10^4	2.5	0.10
Arg-Pro-Arg	2.8×10^5	3.0	1.20

[a]Adapted from Kelly-Wintenberg, K., and T.C. Montie. 1994. Chemotaxis to oligopeptides by *Pseudomonas aeruginosa. Appl. and Env. Microbiol.* **60**:363-367,

[b]10 mM.

[c]Each number represents the average of two separate experiments with duplicate plates from two capillary tubes.

[d]Fold increase over buffer capillary control. A value of 2 is considered significant.

[e]Fold increase or decrease over RTR obtained with arginine. Arginine is considered a strong chemoattractant.

NOTES

PHOTOTAXIS OF *RHODOSPIRILLUM CENTENUM*

One Class Period

Goals

The goals of this experiment are to (1) observe phototaxis of whole colonies of *R. centenum* and (2) to investigate which wavelengths of light are most effective.

INTRODUCTION

Rhodospirillum centenum, ATCC 43720, is a purple, nonsulfur photosynthetic bacterium that was first isolated in 1987 from Thermopolis, Wyoming, hot springs. Like most photosynthetic bacteria, *R. centenum* grows well in an environment that allows it to grow photosynthetically. To this end, it has evolved systems that facilitate accumulation of the cells in light that will support photosynthetic growth. This can be seen microscopically as individual motile cells accumulate in the lighted region of a partially illuminated microscope field. The gathering of cells in illuminated areas results from the fact that cells reverse their direction of swimming upon crossing the boundary between a lighted area and a darkened area. They thus avoid the dark. When *R. centenum* is grown on a solid surface such as agar, a newly discovered phenomenon takes place: The entire colony moves towards the light. This is not simply the avoidance of the dark, as exhibited by individual swimming cells. The movement towards light is called *positive phototaxis*. The type of movement within the moving colonies is called *swarming*, as described next.

During swarming, large groups of cells move in a coordinated manner over the agar via peritrichous flagella. When *R. centenum* is in the swarming phase of its life cycle, colonies that are grown on soft agar plates of rich medium will move rapidly in a straight line from the point of inoculation toward a light source favorable for photosynthesis. More specifically, cells will move toward a light source that emits a relatively high amount of light in the near-infrared region of the spectrum. This corresponds to wavelengths in the range of 800-880 nm, which are absorbed by pigments in the light harvesting and reaction center complexes. Simple tungsten sources (such as household light bulbs) work very well to demonstate this phenomenon of positive phototaxis. Conversely, when colonies of *R. centenum* are subjected to light sources that emit a higher proportion of light in the 580-600 nm region of the spectrum (this includes all fluorescent and quartz-tungsten-halogen sources), the colonies migrate away from the light source. This is called *negative phototaxis*. It is believed that *R. centenum* senses a ratio of available light in these two spectral regions and uses the information to orient itself favorably within the complex, natural environment. It is no coincidence that the two regions of the spectrum that trigger phototactic swarming are also the regions that are biologically active in terms of bacterial photosynthesis. This novel behavior is a graphic and easily visualized example of the principles of positive and negative phototaxis, as well as a general microbial adaptation.

Also, *R. centenum* is ideal for the purpose of demonstrating phototaxis because it possesses another unusual feature: *R. centenum* produces high levels of its photosynthetic reaction center even when grown under aerobic conditions. This is not true for most other photosynthetic bacteria, in which oxygen suppresses assembly of the functional reaction center complex. For *R. centenum*, anaerobiosis is not necessary for growth, for assembly of the reaction center, or for swarming.

MATERIALS

Growth medium
PYVS liquid. Add to 500 ml distilled water:
Bacto-peptone (Difco), 3.0 g
Yeast extract (Difco). 3.0 g
Bacto-soytone (Difco), 4.0 g
vitamin B12 (20 µg/ml stock solution),
1.0 ml
d-biotin (150µg/ml stock solution), 0.1 ml
distilled water to 1000 ml

PYVS culture maintenance plates. Add 15.0 g Sigma Agar to 1000 ml PYVS liquid medium prior to autoclaving. Autoclave and cool to 48°C in a water bath. Swirl container to mix contents. Pour quickly into 15 x 100 mm sterile Petri dishes. After the agar solidifies, invert plates and allow them to stand for at least 36 h at room temperature. One liter should make approximately 40 plates. Extra plates should be wrapped to prevent moisture loss.

PYV swarming plates. Same as the culture maintenance plates except that 8.0 g per liter Sigma agar used. *It is important that plates remain at room temperature at least 24 h before use.* If the agar is too wet, then the colonies are not at a sufficiently high cell density to exhibit phototaxis.

CULTURE MAINTENANCE

R. centenum can be stored for at least several months in stab cultures. Stabs are prepared in PYVS agar (0.8% w/v) that has been dispensed into small test tubes or dram vials. The stabs are inoculated by plunging a loopful of the bacterium into the agar. Allow the culture to grow overnight, illuminated by a standard tungsten bulb. Make sure the culture does not get too hot: *R. centenum* has a growth optimum of 40°C. Remove the culture from the light and store in a refrigerator.

For shorter periods of time, *R. centenum* can be maintained on PYVS plates (agar 1.5% w/v). Streak a loopful onto a fresh plate to obtain cleanly isolated colonies. Incubate the plate at 40°C for 24 h or until colonies are easily visible. (Slightly lower incubation temperatures will work, but plates may need to incubate for 2 days.) *R. centenum* grows as deep red, circular colonies. If the plate is free of contaminants, wrap the edge with Parafilm. Store the plate in the dark at room temperature. Isolated colonies can be picked as needed. Stock maintenance plates should be replaced every week to 10 days.

PREPARATION OF SWARM PLATES

Fill a sterile, screw-capped test tube with sterilized PYVS liquid. The tube should be as full as possible. Pick an isolated *R. centenum* colony from a culture maintenance plate and carefully inoculate the tube. Replace the cap and tighten it firmly onto the tube. The culture should be incubated in the light of a tungsten lamp for 24 to 36 h. Allow the culture to mature.

Inoculate a PYV plate (0.8% agar w/v) with 25 µl of the liquid culture. The liquid should be dispensed as a single spot in the center of the plate. (If micropipettors are not available, a sterile Pasteur pipette may be used to place one small drop of liquid culture on the agar.) Take care not to gouge the agar during this step. Allow the spot to dry completely before attempting to move the plate. Once the spot has dried, place the plate in a 40°C incubator in the dark. Allow the culture to incubate for 24 h or until a single, larger, very dense colony is visible on the plate. Remove the plate from the incubator. Swarm plate cultures must be used for swarming assays immediately after removal from the incubator.

PROCEDURE

Positive phototaxis

Place swarm plate cultures in the light of a single tungsten light bulb (60 watt). Shield the experimental area from ambient light so that only a single light source illuminates the plate. The plates should be placed 25 cm from the light source, and a temperature range of 30-40°C should be maintained. Monitor the plates every hour. The approximate average swarming rate is 10-15 mm per hour. When the swarm has moved to the edge of the place, remove the plate from the light. Plates can be monitored for rates of swarming by measuring the movement of the center of the original colony over time. Filters will be available to test different regions of the spectrum.

Negative phototaxis

Use the same procedure as for positive phototaxis except that the light source is a quartz-halogen lamp or fluorescent light. Allow at least 8 hrs for a good response using a quartz-halogen source (300 watt slide projectors work well) and 24 h for a fluorescent source (cool white tubes are best).

PREPARATION OF DATA

Construct a table that summarizes the behavior of the colonies towards light.

REFERENCES

The procedure for demonstrating phototaxis was contributed by Howard Gest (personal communication).

A discussion of bacterial swarming can be found in Harshey, R. M. 1994. Bees aren't the only ones: Swarming in gram-negative bacteria. *Molec. Microbiol.* **13**:389-394.

NOTES

Experiment 18. Phototaxis of Rhodospirillum centenum

Experiment 19

LIGHT PRODUCTION BY *PHOTOBACTERIUM PHOSPHOREUM*

One Class Period

Goals

This experiment introduces the phenomenon of light production by bacteria and permits the experimenter to analyze the process using the action of inhibitors of cellular processes, provision and absence of metabolites, and the provision of oxygen to whole cells.

INTRODUCTION

A number of prokaryotes produce light. The best studied of these live as mutualistic symbionts with marine animals (vertebrate fishes and squid) in light organs. The intimacy of this symbiosis is reflected in the complex organs elaborated by the animal partner (highly vascularized tissue beds fitted with lenses and eyelid-like skin flaps to control the light emission). Luminous bacteria are shed from these organs and are numerous in pelagic marine waters. Dead animals on beaches can sometimes serve as sites of growth of luminous bacteria, and on moonless nights these are seen to glow in a ghostly fashion.

Luminous bacteria produce light when they simultaneously oxidize riboflavin-5'-phosphate ($FMNH_2$), and a long-chain aldehyde (RCHO), for example, C_{14} (tetradecanal) with oxygen. The reaction is catalyzed by a monooxygenase called *luciferase*, as follows:

$$FMNH_2 + RCHO + O_2 \longrightarrow FMN + H_2O + RCHOOH + light$$

The reason that light is emitted during the reaction is that the flavin becomes electronically excited and subsequently emits light as the electron returns to its ground state. Of the two oxygen atoms in O_2, one becomes reduced to water, while the other is incorporated into the carboxyl group of the carboxylic acid. In order for luminescence to continue, both $FMNH_2$ and RCHO must be regenerated. The aldehyde (RCHO) is regenerated from the carboxylic acid (RCOOH) via reduction by NADPH. The $FMNH_2$ is regenerated by reduction of FMN by NADH. Thus the bioluminescence pathway can be viewed as a shunt leading from the cytochrome respiratory pathway.

Certain eukaryotes such as some algae and fungi, as well as fireflies, are also luminescent, but the reactions are not the same as in prokaryotes. In eukaryotes various compounds called *luciferins* emit light when they become oxidized by oxygen. The luciferins vary depending upon the organism. The luciferin in fireflies is a carboxylic acid, which requires ATP for light emission. The ATP is used to form an AMP derivative of the carboxyl group prior to oxidation.

This experiment tests the effects of respiration inhibitors as well as an uncoupler of oxidative phosphorylation on luminescence of *Photobacterium phosphoreum*.

MATERIALS

Supplies (per team)
-6 16x150 mm test tubes and a rack to hold them
-3 5 ml serological pipettes
-pipette aid to fit serological pipettes
-pipettor and tips to measure 0.05 ml

Equipment
-vortex mixer (can be shared)

Solutions
Culture medium
Photobacterium broth:per liter:(for *Photobacterium* agar add 15 gm powdered agar)

Tryptone	5.0	g
Yeast extract	2.5	g
NaCl	30.0	g
NH_4Cl	0.3	g
$MgSO_4$	0.3	g
$FeCl_3$	0.0l	g
$CaCO_3$	1.0	g
K_2HPO_4	3.0	g
Sodium glycerophosphate	23.5	g

Buffered sea water (SW). Per liter:

NaCl	30.0 g
Na_2HPO_4	2.0 g
KH_2PO_4	1.0 g

5% Sodium β-glycerophosphate

0.05 M KCN (CAUTION! extremely toxic, keep away from acids and dispense to students only in 0.25 ml portions for safety)

0.001 M FCCP (carbonyl cyanide p-(trifluoromethoxy) phenylhydrazone (Sigma C 2920, see reference below) in DMSO. 0.25 ml portion per student. This is a toxic compound.

Cells

6 ml of a suspension of *Photobacterium phosphoreum* ATCC 11040 per student. (See next.)

Preparation of cultures

Make a liter of *Photobacterium* broth. Reserve 800 ml and autoclave the remaining 200 ml. Place 400 ml in each of two 32 oz. prescription bottles and to each add 6.0 g powdered agar; autoclave. Some of this *Photobacterium* agar can be used to prepare Petri plates and slants (2-3 of each). Allow the agar to harden in the bottles resting on their sides.

Streak one of the plates with a *P. phosphoreum* culture and pick a bright (luminous) colony to establish a broth culture in a test tube about a week before the experiment is to be done. A dimly lit room (or an eye patch used to dark adapt an eye, -human eyes can be dark adapted independently), makes picking the bright colony easier. Cloning the culture is necessary since dim mutants accumulate in cultures not so cloned. *Photobacterium* grows best at 18°C, but can be grown from 15 to 25°C.

Two days before the class experiment inoculate each of the agar surfaces in the 32 oz. bottles with 1 ml of the liquid culture and incubate. Just before class wash the cells (which should glow brightly) off the agar with SW and sediment the cells in a centrifuge (3000 x g). Resuspend the cells in 50 ml of SW. The suspension should be glowing (at least at first) quite brightly. Keep this suspension on ice for use in the experiment.

PROCEDURE

To the tubes in a rack make the following additions:

tube 1: 4 ml *Photobacterium* broth, 1 ml cell suspension

tube 2: 4 ml SW, 1 ml cell suspension in SW

tube 3: 4 ml of SW, 0.05 ml of 5% β-glycerophosphate and 1 ml cells

tube 4: 4 ml of *Photobacterium* broth, 0.05 ml of KCN, and 1 ml cells

tube 5: 4 ml of *Photobacterium* broth, 0.05 ml FCCP and 1 ml cells

tube 6: 4 ml of *Photobacterium* broth, 0.05 ml of DMSO and 1 ml of cells

Mix and allow several minutes for the suspensions to come to room temperature.

Darken the room (a dim light or flashlight facili-

tates note taking) and observe the suspensions both immediately after mixing (vortex mixer) and after a short period with no agitation. Which cultures glow most brightly? After a period of rest how does the pattern of light production by different parts of the suspension change? Do any of the suspensions show continuous light production? Do any show none (or little)? Is light production extinguished more rapidly following mixing in one than the others?

Besides these observations you should know that (1) KCN inhibits transport of electrons from reduced cofactors (and, ultimately, from oxidizable substrates) to O_2 via the cytochrome (electron transport) system. (2) FCCP is a weak acid (pKa ca. 7.4) that readily crosses the cytoplasmic membrane of bacteria carrying protons, thereby disrupting the membrane Δp and "uncoupling" ATP synthesis from respiratory electron transport. Uncoupling agents often speed up respiration (O_2 consumption) as a result of their uncoupling action.

PREPARATION OF DATA

Prepare a table showing the behavior of the 6 suspensions (a cartoon may help in one of the columns rather than trying to fit your observations in words in the table) together with the additions made to each of the 6 suspensions. With this table in hand, account for the different behaviors of the different suspensions in a short narrative.

QUESTIONS

1. Does the color of the light produced give a hint as to what excited chemical species is

actually emitting the quantum of light? Hint: are any of the chemical species involved fluorescent, and if so, what color do they fluoresce?

2. How could this experiment be made more quantitative? What additional instrumentation would be needed?

REFERENCES

Haygood, M. G. 1993. Light organ symbiosis in fishes. *Crit. Rev. Microbiol.* **19**:191-216.

Heytler, P. G. 1979. Uncouplers of Oxidative Phosphorylation. In L. V. Fleisher and L. Packer (eds.). *Methods in Enzymology* Vol. LV, pp. 462-472.

Meighen, E. A. 1991. Molecular biology of bacterial bioluminescence. *Microbiol. Rev.* **55**:123-142.

NOTES

APPENDIX A

ANALYSIS OF EXPERIMENTAL DATA

A key stage of any scientific inquiry is to draw a conclusion from the result of an experiment. The misinterpretation of experimental results has been the cause of many past scientific failures and fallacies. In physiology and related sciences the results of experiments are typically numerical, the result of measuring something. The null hypothesis, that is the conclusion that the result of an experiment is due to pure chance at work, or is the result of uncontrolled variables in the experiment, must always be considered and eliminated through careful experimental design and the use of appropriate statistics. Any good experimental writeup should deal with sources of error and should draw what conclusions are called for (and possible) after errors have been discounted.

Measurement errors: Variance, standard deviation, probable error

When presenting quantitative results it is important to give your reader some measure of the error associated with the values you present. Values are typically presented as follows: 0.45 ± 0.05. The number following the $\pm$ may indicate the range of values averaged to get the 0.45, or it may be some statistical measure of the uncertainty associated with the 0.45 value. It is important to indicate what this uncertainty means and how it was derived.

Types of errors and their causes
Errors may be mistakes, systematic or random in nature. Mistakes are failures to perform properly in recording or transcribing data or in performing calculations. Mistakes can be avoided by having another check your work or by redoing and checking it yourself. Systematic error results from bias built into the process of gathering data. Possible sources of systematic error occur when using improperly calibrated equipment or when incorrectly but consistently misusing measuring equipment. The use of standards and proper controls helps to eliminate systematic errors. Random errors are unavoidable. They result from normal (given a certain experimental design) and proper use of well-calibrated apparatus and reflect the inevitable occurrence of "noise" in the measurement process. An example of random error might be the errors introduced in reading a value given by a pointer moving in front of a scale. Parallax errors in reading the scale resulting from normal changes in the position of the observer's head and eyes during separate readings of the scale would be largely random. While design changes (e.g., adoption of a mirror scale) may help to reduce such error, it is often sufficient to simply repeat a measurement a number of times and to take an average of the values to reduce greatly the effects of such random error. Making a simple scatter plot of the readings (e.g., reading value vs. order of reading) will sometimes be very helpful since outlier values become conspicuous and trends (e.g., as when readings are taken while the measurement apparatus is warming up, producing drift) are revealed clearly this way.

How error is expressed

To illustrate measurement error and how it might be expressed, consider a situation where ten absorbance readings are taken of duplicate samples.

The range of readings is from 0.629 to 0.670. The arithmetic mean (average) of these readings is 0.647. If one expressed the uncertainty of these values by giving the range, then the average reading would be 0.647 ± (0.670-0.629) or 0.647±0.041. This is the most conservative way of expressing the uncertainty in this group of determinations. The difference between each number in the set and the mean is its deviation. If one squares all of the deviations (variations) and takes their arithmetic average, one has calculated the variance, that is, the mean of the squares of the variations from the mean ("mean square of the deviation"). The square root of the variance is called the standard deviation, also called the "root mean square of the deviation". The symbol for the standard deviation is s or σ and is a good measure of the "spread" of the values. For these absorbance readings, the standard deviation, $s = 0.013$. Approximately 2/3 of the measurements will fall within one standard deviation of the mean. A more intuitive way of expressing the spread, and one that makes fewer assumptions about sources of error, sample size, etc., is probable error (P.E.). For small samples P.E. can simply be estimated since it is a ± value on either side of the mean that will cover 1/2 of the measured values. The probable error for the absorbance readings is 0.017 , that is, the number can be expressed as 0.647±0.017. If a large sample is available and a normal distribution is shown, the P.E.=0.674s or P.E.=0.845 the "mean deviation from the mean".

Significant numbers: A convention

Consider the number 1.8×10^{-1} or 0.18 mg/ml. The last significant digit is the second digit after the decimal point, and it cannot be known more precisely than ± 0.005. Therefore, the number falls between 0.175 and 0.185 or, stated more precisely, is $1.8 \times 10^{-1} \pm 5 \times 10^{-3}$ mg/ml. Another example is 1.752 mg/ml. The last significant number is the third digit after the decimal point, and therefore the number cannot be known more precisely than ± 0.0005. Thus, the number stated precisely is $1.752 \pm 5 \times 10^{-4}$ mg/ml, or anywhere between 1.7515 and 1.7525 mg/ml.

Rounding off numbers

It is usually necessary to round off numbers when presenting data so that the data do not contain more significant digits than the least precise measurement. For example, when you use a standard curve to calculate the amount of protein or other material in your sample, you cannot be more precise than the precision with which the standard was measured. Suppose you used your protein standard curve to calculate a protein concentration in your sample. If the standard protein solution was precise to the second digit after the decimal point (e.g., 1.00 mg/ml), then you must round off the numbers obtained for your samples to the same precision. For example, 0.3542 mg/ml would be rounded off to 0.35 mg/ml.

Precision versus accuracy

Although these two words are used interchangeably in ordinary speech and writing, they have specific technical meanings useful in the description of errors. Precision refers to the closeness of multiple determinations; accuracy refers to degree of agreement with the true value. Note that high precision does not mean that a group of measurements of the same quantity are likely to be accurate. Nor can one know whether a group of imprecise measurements is inaccurate.

Propagation of errors

The combination of two or more measurements in one calculation to obtain a final result can result in propagation of error(s). The relative effects of error propagation are greatest when the two numbers resulting from measurements are combined; least when multiplied by one another or subtracted. The extent of the propagation effect can be estimated most simply by doing the calculation once with both values at the mean plus the highest values of P.E. expected for both and, again, using the mean plus the lowest values of P.E.

Coefficient of regression

When graphing values to obtain, for instance, a standard curve that should be linear, one may obtain a measure of the error by giving the coefficient of regression (R^2) (also r^2 or r value). A perfectly fit straight line should have an r^2 of 1.0. Values less than this indicate a relatively poor fit.

Graphs are typically designed to show how two or more experimental variables vary in relation to one another. Known errors or reproducibility can be indicated in such plots by error bars that indicate an expected or measured range of error. R-values (r^2) are usually calculated automatically as a feature of the software in most graphing packages for computer use and are, of course, only useful if the relationship between the variables in a plot is known or may be assumed. Simple graphing and some column statistics are typically available through spreadsheets as well.

Volumetric accuracy

When using less than the total volume of a measuring container, the error increases. Therefore, if accuracy is of the utmost importance when making a quantitative transfer (e.g., with a pipette or a graduated cylinder), always use most or all of the volume of the measuring device for best accuracy. For example, it is more accurate to use a 1 ml pipette to measure 1 ml then to use a 10 ml pipette to measure 1 ml. By the same token, it is more accurate to use a 10 μl pipetter to measure out 1 μl rather than to use a 1000 μl pipetter, and it is more accurate to use a 10 or 25 ml graduated cylinder to measure 10 ml rather than a 1000 ml graduated cylinder. This is especially important when doing a dilution series. Why?

Diluting a concentrated solution may be more accurate than weighing out small quantities. Weighing out less than 1 mg on a classroom balance cannot be done with great accuracy because the balance is not accurate to the second decimal place. By the same token, weighing out less than 0.5 g is not very accurate on a nonanalytical balance. It is therefore more accurate to weigh out larger amounts and do a dilution or weigh out larger amounts and dissolve the material in a larger volume.

HOW TO DO SIMPLE DILUTIONS

Suppose you wanted to dilute a solution or a suspension of bacteria. How would you do it? To begin with, you need to make certain that the units of the concentrated material are the same as the dilute material. For example, it might be g/ml or cells/ml. You can use:

$$C_1 \times V_1 = C_2 \times V_2 \qquad (1)$$

In this equation 1, C_1 and V_1 are the initial concentration and the volume to be transferred, respectively, and C_2 and V_2 are the final concentration and the final volume, respectively. What Eq. 1 tells us is that the total amount of solute (or cells) that is transferred (C_1 x V_1) must equal the total amount in the final solution or suspension (C_2 x V_2). In Eq. 1, it is important to understand that V_2 is equal to the volume of the diluent plus the volume of the sample that is transferred to it. (The diluent is the liquid used for the dilution. It is frequently water, buffer, or growth medium.) For example, if the diluent was 9.5 ml and 0.5 ml was transferred, then the final volume, V_2, is equal to 10 ml. Sometimes the volume of sample transferred is insignificant compared to the volume of the diluent and can be ignored. For example, if the volume of sample transferred to the diluent is less than 5% of the volume of the diluent, it is frequently ignored, unless one is performing a serial dilution.

Suppose you wanted to dilute a solution from 1.2 g/ml to 6 x 10^{-2} g/ml, and the final volume was 10 ml. Then you can use Eq. 1 and solve for V_1. Thus,

$$V_1 = (6 \times 10^{-2} \text{ g/ml} \times 10 \text{ ml}) / 1.2 \text{ g/ml} = 0.5 \text{ ml}$$

You would therefore transfer 0.5 ml of the concentrated solution to 9.5 ml of diluent. Suppose you were asked to dilute 1.2 g/ml into 10 ml of water to a final concentration of 6 x 10^{-2} g/ml. Then, you can use Eq. 1 but substitute 10 + V_1 for the final volume, that is, V_2. In this case, you would transfer 0.52 ml to 10 ml.

Dilution terminology: Fold dilutions (D) and dilutions (1/D)
One can state dilutions in terms of the increase in volume associated with the dilution, called the fold dilution (D). Suppose you are asked to make a 20-fold dilution. This means that when you make the dilution, you increase the volume of the starting material 20-fold, for example, 1 ml to 20 ml, with diluent. Once diluted, the concentration of the material is often said to be 1/D of its original concentration. That is to say, the dilution is simply the reciprocal of the fold dilution. In this case, it would be 1/20 or 5 x 10^{-2} of its original concentration. We will call 1/D the dilution and D the fold dilution. For example, a 2 x 10^{-3} dilution is a 5 x 10^2-fold dilution. If a 2 x 10^2-fold dilution (D) yielded 100 colonies per unit volume in a viable count, then the concentrated suspension has 100 x D or 2 x 10^4 viable cells per unit volume. One can also divide by the dilution (1/D) to get the same figure: 100 ÷ (1/D) = 100 x D.

Sometimes the fold dilution must be calculated from the starting and final cell densities or concentrations of solute. In this case, one uses:

$$\text{Fold dilution} = \text{D} = \text{concentrated/dilute} \quad (2)$$

For example, if you were to dilute a cell suspension from 2 x 10^8 cells per ml to 3 x 10^6 cells per ml, this would be a (2 x 10^8)/(3 x 10^6) or a 67-fold dilution.

Once D is known, you can use the following equation to dilute a solution or cell suspension a certain amount:

$$y(D) = x + y \quad (3)$$

where y can be the ml of concentrated solution added to x ml of diluent, D is the fold dilution, and x + y is the final volume. Suppose you were asked to dilute a glucose solution 20-fold into 10 ml of water. How much of the concentrated glucose solution would you transfer?

$D = 20$

$y \; ml \; (20) = 10 \; ml + y \; ml$

$y = 10/19 = 0.53 \; ml$

How to do serial dilutions

For a serial dilution, the final dilution is the product of the intermediate dilutions. For example, a 2.7×10^3-fold dilution (D) can be obtained using any of the following fold dilutions:

$(10^3)(2.7)$

$(10^2) \, (10^1)(2.7)$

$(10^1)(10^1)(10^1)(2.7)$

You must decide which dilution series you will use. For example, suppose you wanted to do the $(10^2)(10^1)(2.7)$-fold dilution. Then you need three tubes. The first tube will be diluted 10^2-fold (a 10^{-2} dilution). The second tube will be a 10^1-fold dilution of the first tube so that the total fold dilution is $10^2 \times 10^1$ or 10^3. This would be a 10^{-3} dilution. The third tube would be a 2.7-fold dilution of the second tube or a total dilution of 2.7×10^3, which would be a 3.7×10^{-4} dilution. You should label your tubes 10^{-2}, 10^{-3}, and 3.7×10^{-4}. The order of the tubes can be changed. However, it is often convenient to have the largest dilutions first if the amount of starting material is limited. Now you can calculate how much material to transfer for your serial dilutions. Suppose the final volume in each tube will be 10 ml. You can use Eq. 3 , where y is the number of ml to be transferred and 10 ml is the final volume. Thus $y = 10/D$. Then, for the 10^2-fold dilution, transfer $10/10^2 = y = 0.1$ ml. Since the final volume is 10 ml, the 0.1 ml must be transferred into 9.9 ml

(10 - 0.1) of water. For the 10^1-fold dilution, transfer $10/10^1 = 1$ ml into 9 ml (10-1) of water. For the 2.7-fold dilution, transfer $10/2.7 = 3.7$ ml into 6.3 ml (10 - 3.7) of water. See the problems in Appendix F.

GUIDELINES FOR WRITING LABORATORY REPORTS

Each report should have five sections: Abstract, Introduction, Methods, Results, Discussion. If the laboratory exercise is very simple, then the Results and Discussion sections can be combined. Always use the past tense when referring to your work or to the results. You can use the present tense when referring to the work of others or to common knowledge. Use the first person active or third person passive voice. Examples: "The isolation buffer was added to the cell lysate" (third person passive). "I added the isolation buffer to the cell lysate" (first person active). Examine any of the scientific journals in the library for the general form and style of a scientific paper.

Abstract

The Abstract should be one paragraph succinctly summarizing the main results and conclusions. The purpose of the abstract is to inform the reader of the main findings of the research. To that end, the abstract should state precisely what was measured and the key numbers that were obtained.

Introduction

The Introduction should describe the overall experiment, its physiological significance, and any other introductory information about the experiment that you feel would be of interest to the reader. For example, if the purpose of the laboratory is to purify an enzyme and to measure its kinetic constants, then the Introduction should describe the enzyme, its physiological role, and the significance of the K_m and V_{max}. Generally, the Introduction will be very short, no longer than half or three-fourths of a page.

Methods

The Methods section should describe the procedures that were followed, but it should not be a duplication of sections of the laboratory manual. For example, if the enzyme was purified by Sephacryl chromatography, then simply state so, without describing how the column was constructed. If you measured the RNA content of bacteria after trichloroacetic acid (TCA) extraction, then simply state: "RNA was extracted using hot TCA extraction and assayed using the orcinol procedure." Protein assays are simply referred to as the Lowry assay without writing: "Four ml of solution C were mixed with 1 ml of sample." In other words, the Methods section should have sufficient detail so that it simply identifies for the reader the procedures that were followed, so long as these procedures are described in detail in the laboratory manual.

Results

The Results section should have results only, no methods or discussion, and is usually as long or longer than the Discussion section. For example, if you are describing experiments that measured the amounts of RNA in the cells at different growth rates, then simply describe the results of the experiment. Always refer to a figure or table that contains the results. However, simply referring to the figure or table is not enough. In other words, do not simply state, "The results are shown in Table 1 and figure 1" and then move on to the Discussion. You must also summarize for the reader the important results that are in the tables and figures. For example: "As seen in Table 1 and Figure 1, the RNA content increased twofold with a tripling of the growth rate. Protein, on the other hand, increased by only 50% (Table 1, Fig. 2)." However, do not discuss the expected results or the significance of the results in the Results section. That discussion is reserved for the Discussion section.

Discussion

The Discussion section should include a recapitulation of your own results, the class results, a comparison of the two, a discussion of discrepancies (if any), and a discussion of the physiological significance of the results. The Results and Discussion sections can be combined.

How to make a table

Each table should have a table number and a title. The table should be organized into vertical columns, each having its own heading. That is to say, tables are organized so that all like elements are organized into vertical columns. There are no vertical lines. Units should be in the headings. The table should be self-explanatory without a need to refer to the text. Tables can have footnotes to explain symbols or to impart any additional information that would be useful to the reader.

Table 1 Rates of oxygen uptake and final cell yield as a function of carbon source[a]

Carbon source	Oxygen uptake (μmoles/min)	Cell yield (A_{650})
glucose	2.4	0.95
succinate	5.7	1.71
citrate	0.2	0.04
malate	4.8	1.72

[a]Cells were grown at 37°C on a rotary shaker in minimal salts media supplemented with the indicated carbon source. The initial A_{650} was between 0.03 and 0.06.

Table 2 Effect of temperature and pantothenate on growth of *pur* mutants[a]

Strain	Specific growth rate (μ) (doubling h[-1]) and final yield (A_{650})							
	37°C				30°C			
	- pantothenate		+pantothenate		- pantothenate		+ pantothenate	
	μ	A_{650}	μ	A_{650}	μ	A_{650}	μ	A_{650}
JW 135	0.04	0.10	0.08	0.10	0.05	0.10	0.04	0.12
EW 348	0.05	0.10	0.03	0.12	0.07	0.13	0.06	0.10
SW 137	0.06	0.14	0.08	0.53	0.17	0.61	0.18	0.94

[a]Strains were grown in minimal media with or without pantothenate. Initial A_{650} was between 0.02 and 0.04.

Figures and legends

Make certain that each figure has a legend. The legend should have the figure number, the title of the figure, and an explanation of what the symbols in the figure mean. For example, if there is more than one line in the figure each line should have different symbols (e.g., open and closed squares) and these symbols should be explained in the legend (not in the figure). The purpose of the legend is to make the figure understandable without recourse to the text. Use linear graph paper when plotting standard curves and any other functions that show linear relationships. Use semilog paper only when plotting growth or some physiological response that changes exponentially with time or some other variable.

REFERENCES

The following books are good references for statistical analysis of data. The one by Wilson is biochemically oriented and is full of practical pointers. The book by Beveridge is more biologically oriented and deals with the role of imagination and how human factors affect the practice of science.

Beveridge, W. I. B. 1973. *The Art of Scientific Investigation.* Vintage Books, 3rd edition.

Wilson, Jr., E. B. 1952. *An Introduction to Scientific Research.* McGraw-Hill Book Co., Inc.

Software is available for graphing and statistical analysis. Some examples are:

ChemWindow (ChemIntosh)(SoftShell International) - sophisticated drawing programs for chemical structures and reactions.

Cricket Graph - a graphing and drawing program for Macintosh.

SigmaPlot (Jandel Scientific) - a more sophisticated graphing program for Mac or PC including Windows.

SigmaStat (Jandel Scientific) - a statistics program for Mac or PC (Windows) with expert systems ("Advisor") features.

APPENDIX B

LABORATORY SUPPLIES AND EQUIPMENT

The supplies per team of students are given in the description of the individual experiments. In general, however, the laboratory should be equipped with standard supplies, including the following:

analytical and nonanalytical balances
eye protectors
a fume hood
refrigerated centrifuge with appropriate rotors and centrifuge tubes
water baths
Bunsen burners
colorimeters or spectrophotometers and appropriate tubes

vortex mixers
pH meter
marking pens
latex gloves
0.1, 1, 5, and 10 ml sterile and nonsterile pipettes
20, 200, and 1000 μl pipetters with tips
pipette bulbs
Pasteur pipettes
16 x 150 mm test tubes. These are convenient to use on a routine basis. This is because when these test tubes contain 5-10 ml of liquid (which is usual for most of the experiments described in this manual), they can vortexed without spilling the material.
test tube racks
distilled water wash bottles
37^0C incubator

APPENDIX C

MEDIA

Standard mineral base (SMB*)
Solution A

    0.2 M solutions of Na_2HPO_4 and KH_2PO_4 mixed to give a desired pH. (See phosphate buffer, below.)

Solution B

    $MgSO_4$ -$7H_2O$, 20.0 g
    $CaCl_2$, 1.0 g
    distilled water to 1000 ml

Ferric-EDTA

    Dissolve 17.9 g sodium-EDTA and 3.23 g KOH in 186 ml distilled water. Dissolve 13.7 g $FeSO_4$ -$7H_2O$ in 364 ml distilled water.
    Mix the two solutions, and bubble with air overnight to oxidize the Fe^{2+} to Fe^{3+}. Store in a dark place.

Solution C

    Ferric-EDTA, 10 ml
    $ZnSO_4$-$7H_2O$, 50 mg
    $MnSO_4$-H_2O, 50 mg
    $CuSO_4$, 10 mg
    cobalt nitrate, 10 mg
    sodium borate, 10 mg
    sodium molybdate, 200 mg
    distilled water to 100 ml

10% $(NH_4)_2SO_4$ (sometimes NH_4Cl is used instead)

Mix 100 ml Solution A, 10 ml of Solution B, 1 ml of Solution C, 10 ml of 10% $(NH_4)_2SO_4$ or NH_4Cl, and adjust the volume to one liter with distilled water.

M-9 medium
    Na_2HPO_4, 7.0 g
    KH_2PO_4, 3.0 g
    NH_4Cl, 1.0 g
    NaCl (25%), 2.0 ml
    $FeCl_3$ (0.01 M), 0.3 ml
    water to 1000 ml (usually tapwater is used.)

BUFFERS

phosphate buffer

Solution A
0.2 M KH_2PO_4

Solution B
0.2 M Na_2HPO_4

A (ml)	B (ml)	pH
93.5	6.5	5.7
87.7	12.3	6.0
77.5	22.5	6.3
51.0	49.0	6.8
39.0	61.0	7.0
28.0	72.0	7.2
16.0	84.0	7.5
8.5	91.5	7.8
5.3	94.7	8.0

Tris (hydroxymethyl)aminomethane (Tris) buffer
Solution A. 0.2 M
Tris(hydroxymethyl)aminomethane
(24.2 g in 1000 ml)
Solution B. 0.2 M HCl

Add x ml of B to 50 ml of A and dilute to 200 ml.

x	pH
5.0	9.0
8.1	8.8
12.2	8.6
16.5	8.4
21.9	8.2
26.8	8.0
32.5	8.0
38.4	7.6
41.4	7.4
44.2	7.2

SOLUTIONS

phosphate-buffered saline (PBS) (10X)
Na_2HPO_4, 12.36 g
$NaH_2PO_4 \cdot H_2O$, 1.80 g
NaCl, 85.00 g
water to 1000 ml

This is a 0.1M solution. When diluted 1:10 it is a 0.01M solution, pH 7.6

acid alcohol
HCl, concentrated, 3 ml
95% ethyl alcohol, 96.0 ml

Benedict's reagent
sodium citrate, 17.3 g
Na_2CO_3 anhydrous, 10.0 g
$CuSO_4 \cdot 5H_2O$, 1.73 g
distilled water to 100 ml

Dissolve sodium citrate and sodium carbonate by heating in about 60 ml of water. Cool solution and make up to a volume of 85 ml with water. Dissolve copper sulfate in about 10 ml of water and add slowly, with constant stirring, to the citrate-carbonate solution.

APPENDIX D

RNA ASSAY USING ORCINOL

Acidified orcinol reacts with aldopentoses to yield a green color. The reaction is used to measure ribose in RNA. DNA reacts with approximately 20% of the color of RNA. Use caution when using concentrated hydrochloric acid (HCl). Avoid breathing the fumes and exposure to skin and eyes. When using concentrated HCl work in a fume hood.

Reagents

stock orcinol (store in a dark bottle; if the orcinol crystals are yellow, then recrystallization is necessary):

>   6.0 g orcinol
>   100 ml ethanol

stock $FeCl_3$:

>   100 mg $FeCl_3$-$6H_2O$
>   100 ml conc. HCl

standard:

>   1.00 mg/ml RNA dissolved in 5%
>   TCA. Heat at 70°C for 30 min
>   to dissolve.

orcinol reagent (prepare on day of assay; the complete reagent should be bright yellow):

>   100 ml ethanolic orcinol
>   100 ml stock $FeCl_3$

Procedure

1. Set up a series of 5 test tubes for the standard curve and 2 test tubes for each volume of sample to be assayed. The sample should be assayed in duplicate. Make certain that you include a tube for the reagent blank which will include everything but RNA.

2. Add sample or standard and water to 1 ml. The standards should contain from 10 to100μg of RNA.

3. Add 2 ml of orcinol reagent. Mix and cap each tube with a marble. Place in a 90 to 100°C water bath for 30 min. A green color should develop. Cool and read at 665 nm.

The temperature of incubation is very important. If the temperature falls to 80°C, then the samples must be incubated for 90 min.

REFERENCES

Daniels, L, Hanson, R. S., and Phillips, J. A. 1994. Chemical analysis, pp. 512-554. In: *Methods for General and Molecular Biology*. Gerhardt, P., Murray, R. G. E., Wood, W. A., and Krieg, N. R, (eds.). American Society for Microbiology. Washington, D.C.

LOWRY PROTEIN ASSAY

In the Lowry assay a blue color develops because peptide bonds react with the alkaline copper solution (the biuret reaction) and phosphomolybdate-phosphotungstic acid in the Folin reagent is reduced by tyrosine residues in the protein. The reduction of the phosphomolybdate-phosphotungstic acid accounts for most of the color.

Reagents

Reagent A.
    sodium carbonate, 20 g
    sodium hydroxide, 4 g
    sodium potassium tartrate, 1.6 g
    sodium dodecyl sulfate (SDS), 10 g
    water, 1 liter

SDS will precipitate in the cold. Make certain that there is none at the bottom of the bottle before using it. You can dissolve SDS by placing reagent A in a 30°C water bath.)

reagent B.
    $CuSO_4 \cdot 3H_2O$, 400 mg
    water, 10 ml

reagent C.
    100 ml reagent A to which 1 ml reagent B has been added while mixing. Will need 50 ml for each assay which includes the standard curve plus samples. This is good for about 1 day.

Folin & Ciocalteau's phenol reagent
    dilute 1:1 with water on day of use.
    Will need 5ml for each assay.

standard protein (bovine serum albumin)
    1 mg/ml in water(1 µg/µl) Will need 0.6 ml for each assay.   It can be kept in the refrigerator, If it is stored frozen, the defrosted solution should be well mixed before using.

Procedure

1. Make a series of 5 16 x 150 mm test tubes (or 13 x 100 mm test tubes) into which the standard bovine serum albumin will be placed, 1 test tube for a reagent blank (everything but the protein), and 2 test tubes for each volume of sample to be assayed. (You will assay your samples in duplicate.)

2. Add the standard bovine serum albumin over a 20 to 100 µg range i.e., from 20 to 100 µl.

3. Add 20 to 400 µl of sample such that the amount of protein delivered is between 20 and 100 µg.

4. Add 4 ml of fresh reagent C to each of the test tubes and mix. (The blank tube has just reagent C.)

5. Incubate 15 to 30 min at room temperature.

6. Add 400 µl of Folin-phenol reagent:water (1:1) to all the tubes. Mix immediately after adding the reagent to each tube. You can mix by either flicking the tube with your finger or by vortexing.

7. Incuate at least 45 min at room temperature. The color is stable.

8. Read absorbance at 660 nm.

Draw a standard curve. The abscissa (x axis) should have the total weight of protein in the tube in micrograms (not the concentration) and the ordinate (y axis) should show the absorbance.

The Lowry assay is a convenient and simple protein assay. However, it is subject to interference by potassium ion, magnesium ion, EDTA, Tris, thiol reagents, and carbohydrates. such as sucrose. An alternative to the Lowry assay, i.e., the Bradford assay, is described next.

REFERENCES

Lowry, O. H. Rosebrough, N. J., Farr, A. L., and Randall, R. J. 1951. Protein measurement with the Folin phenol reagent. *J. Biol. Chem.* **193**:265-275.

BRADFORD PROTEIN ASSAY

Coomassie Brilliant Blue G-250 is red in solution but turns blue when it binds to protein. The Bradford assay measures the protein-dye complex at 595 nM.

Reagents

Bradford reagent (0.01% [w/v] Coomassie Brilliant Blue G-250, 4.7% [w/v] ethanol, 8.5% [w/v] phosphoric acid)

> Dissolve 100 mg of Coomassie Brilliant Blue G250 in 50 ml 95% ethanol.

> Add 100 ml of 85% (w/v) phosphoric acid.

> Dilute to 1 liter with water.

The Bradford reagent is also available commerically, e.g., from Sigma Chemical Co. St. Louis, MO.

Procedure

1. To a series of 13 x 100 mm test tubes add standard BSA over a range of 10 to 100 µg of protein e.g., 10, 20, 40, 80 100 µg.

2. Add sample (between 10 and 100 µg) in duplicate tubes.

3. Add water to all tubes to a final volume of 0.5 ml. Include a tube to which protein has not been added. You will blank the spectrophotometer with this tube.

4. Add 5 ml of Bradford reagent to all the tubes and mix immediately. Mix either by vortexing or by inverting the tube after covering the mouth of the tube with parafilm.

5. After 5 min. read the absorbance at 595 mM. The color is stable for at least 1 hour.

REFERENCES

Bradford, M. M. 1976. A rapid and sensitive method for the quantitation of microgram quantities of protein utilizing the principle of protein-dye binding. *Analyt. Bioch.* **72**:248-254.

ASSAY FOR α-AMYLASE USING STARCH CONJUGATED TO A DYE

The assay monitors the α-amylase-induced solubilization of Remazobrilliant Blue R (RBB) (Sigma Chemical Co.), which is covalently bound to starch. It should be pointed out that this assay is not as sensitive as the iodine assay described in Experiment 8.

Reagents

A suspension of RBB-starch (2%) in 0.02 M sodium phosphate buffer (pH 7.0).

Procedure

1. Place 22.5 ml of the reagent in a 125 ml Erlenmeyer.

2. Add 2.5 ml of enzyme.

3. Cap and incubate on a rotary shaker in a 37°C water bath.

4. At 0, 10, 20, and 40 min remove 5 ml and add to 2 ml dilute (6N) acetic acid (reducing pH to 4).

5. Filter with Whatman #1 filter paper and read filtrate at 595 nm. The blank is a similar reaction mixture that omits enzyme.

REFERENCES

Rinderknecht, H., Wilding, P., and Haverbac, B. J. 1967. A new method for the determination of α-amylase. *Experientia.* **23**:805.

APPENDIX E

THE BEER-LAMBERT LAW

The Beer-Lambert law (Eq. 1) relates the absorption of monochromatic light to the concentration in moles per liter (c) of the substance absorbing the light, the length of the light path in centimeters through the sample (l), and the molar extinction coefficient (e).

$$A = \text{absorbance (or optical density)} \qquad (1)$$
$$= \log_{10} I_o / I = ecl$$

In Eq. 1, the incident light intensity is I_o and the transmitted light intensity is I.

The Beer-Lambert law also applies to light scattering (turbidity measurements). The turbidity of a suspension of cells is equal to cl, where c is the mass of cells and l is the light path. When the cell density is too high, then some of the light is rescattered and directed toward the phototube, thus lowering the absorbance reading. See Experiment 1 and Fig. E.1.

REFERENCES

For a discussion of the Beer-Lambert law and turbidity measurements, see Koch, A. 1994. Growth measurement. pp. 248-277. In: *Methods for General and Molecular Bacteriology*. Gerhardt, P., Murray, R.G.E., Wood, W. A., and Krieg, N. R. (eds.). American Society for Microbiology, Washington, D.C.

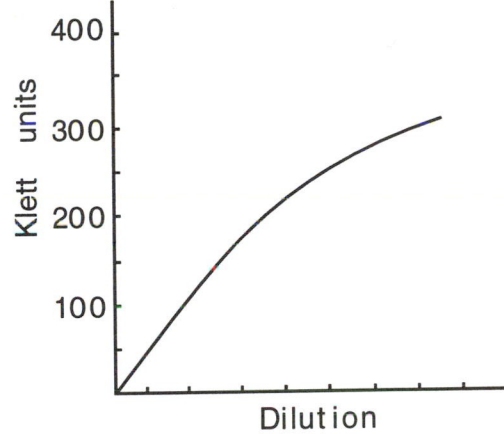

Fig. E.1 Klett units (red filter) versus dilution of a *Proteus vulgaris* culture. Cells were centrifuged and resuspended in water for the measurements.

Table E.1 Relative absorbances of a bacterial cell suspension measured at different wavelengths[a]

A_{660}	A_{540}	Klett units (#66 red filter)	Klett units (#54 green filter)
x	1.3 x	124 x	248 x

[a]The ratios apply only at a sufficiently low turbidity so that the Beer-Lambert law applies. Measurements were made with cell suspensions of *E. coli*. The absorbance readings will vary with the colorimeter or spectrophotometer. The ratios in the table are only approximations.

RELATIVE ABSORBANCIES

Usually, the cell density of a bacterial culture is measured using blue, green, or red light. The absorbance readings (due to light scattering) are higher at the lower wavelengths. Table E1 gives an idea of how the absorbances vary with wavelength, and also the relationship of absorbance to Klett units.

CELL DENSITY versus KLETT UNITS

Approximately 1.5×10^7/ml *E. coli* cells will have a turbidity of 1 Klett unit (red filter).

UNITS

$1\ g = 10^3\ mg = 10^6\ \mu g = 10^9\ ng = 10^{12}\ pg = 10^{15}\ fg$

$1\ l = 10^3\ ml = 10^6\ \mu l$

APPENDIX F

QUANTITATIVE PROBLEMS

(See Appendix G for answers.)

1. How many ml of a culture that is 5.0×10^8/ml should you transfer to 40 ml of fresh medium in order to obtain a density of 3.5×10^6/ml?

2. Suppose you diluted a culture of cells 2500-fold that was initially 5.0×10^8/ml. What is the fold dilution, what is the dilution, and what would be the final cell density?

3. Assume that you wished to dilute a sample 6×10^5 fold. How would you do this using serial dilutions?

4. Mercaptoethanol is sold as a 14.3 M solution. How many microliters would you add to 25 ml to make a 50 mM solution?

5. The molecular weight of adenosine monophosphate (AMP) is 347.2. How much would you dissolve in 50 ml to make a 25 mM solution?

6. Concentrated hydrochloric acid is approximately 18 M. How many ml of concentrated hydrochloric acid are added to make a 100 ml (final volume) solution that is 2M?

7. Suppose you assayed for the enzyme ß-galactosidase and the rate was A_{420}/min = 0.01. 50 µl of enzyme were added to the sample tube. The concentration of protein in the stock solution was 2 mg/ml. What is the specific activity?

8. The absorbancy (A) of a solution follows the Beer-Lambert Law, that is, $A = ecl$, where e is the molar extinction coefficient (liter mole^{-1} cm^{-1}), c is the concentration of the absorbing substance in moles per liter, and l is the light path in centimeters. The molar extinction coefficient for NADPH in a cuvette with a 1 cm light path is 6.22×10^3. What is the concentration of NADPH if the absorbancy is 0.2?

9. Assume you assayed for the enzyme glucose-6-phosphate dehydrogenase by measuring the increase in A_{340} as NADPH was being produced in a volume of 3 ml in a cuvette with a light path of 1 cm. The mM extinction coefficient for NADPH is 6.22 liter mmole^{-1} cm^{-1}. (See Problem 8.) The increase in A_{340} was 0.01/min. A unit of enzyme (U) is defined as that amount of enzyme that will catalyze the transformation of one micromole of substrate per min. How many units of enzyme were there?

10. Assume you measured the absorbancy of a 29 mg/liter solution of uridine-5'-triphosphate (UTP) in a cuvette with a 1 cm light path and obtained a value of 0.500. The molecular weight of UTP is 586. What is the molar extinction coefficient?

APPENDIX G

SOLUTIONS TO PROBLEMS IN APPENDIX F

There are two ways to solve dilution problems. You can use either eq. 1 or eqs. 2 and 3.

$$C_1V_1 = C_2V_2, \quad (1)$$

where C_1 is the concentration of the solution or bacterial suspension to be diluted, V_1 is the volume to be transferred, C_2 is the final concentration, and V_2 is the final volume. Note that V_2 is equal to V_1 plus the volume to which V_1 is transferred.

$$\text{concentrated} / \text{dilute} = D \quad (2)$$

D is the fold dilution. (1/D is the dilution.)

$$y(D) = V + y \quad (3)$$

where y is the volume to be transferred and V + y is equal to the final volume.

1. Using eq. 1: $(5.0 \times 10^8/ml)(V_1) = (0.035 \times 10^8/ml)(40 + V_1)$. Solve for V_1.

   Ans. 0.29 ml.

2. Ans. Fold dilution is 2500. dilution is 4.0×10^{-4}. Final cell density is 2×10^5 cells/ml.

3. First set up a series of practical dilutions for example where each dilution is betwen 2- and 100 fold. For example, they could be mostly 10 fold dilutions. The product of the dilutions should equal 6×10^5. Start with the great-est dilution and proceed toward the smallest. Label your tubes 1/D. Decide on the final volumes in your tubes (10 ml is convenient) and use Eq. 3 to find the volumes (y) to be transferred.

4. First find the fold dilution, D. This is $14.3 \div 0.05$, i.e. 286. Then use Eq. 3: $y(D) = 25 + y$, where y is the number of ml of mercptoethanol to be added to 25 ml. Because the dilution, 286, is so large, y can be ignored in the right-hand side of the equation. You can also use Eq. 1.

   Ans. 0.0874 ml or 87.4 μl

5. A 25 mM solution is 25 mmoles per liter. This is 50 (25/1000) = 1.25 mmoles per 50 ml. Since the molecular weight is 347.2, 347.2 mg = 1 mmole.

   Ans. 1.25×347.2 or 434 mg

6. First find the fold dilution, D, using Eq. 2. This is $18 \div 2 = 9$. Then use Eq. 3 i.e. 9Y = 100. Therefore, 11.1 ml should be added to 88.9 ml to make 100 ml. If the concentrated HCl were added to 100 ml, then 9Y = 100 + Y. Alternatively, you can use Eq. 1, where $C_1 = 18$ M, $C_2 = 2$ M, $V_2 = $ the final volume, and V_1 is the volume transferred.

   Ans. 12.5 ml if the concentrated acid were added to 100 ml of water.

7. The specific activity is the rate of enzyme activity divided by the mg of protein in the

assay tube. Since 50 μl of a 2 mg/ml solution of protein were added, the total amount of protein in the assay tube was 0.05 x 2 = 0.1 mg.

Ans. 0.1 absorbance units/min per mg protein. If the enzyme assay was performed with whole cells rather than a cell-free extract, then you can determine the specific activity per dry weight of cells, per turbidity unit, or per specified amount of cells.

8. Use the equation A = ecl. That is,

$$0.2 = \frac{6.22 \times 10^3 \text{ liters} \cdot c \text{ moles} \cdot 1 \text{ cm}}{\text{moles} \cdot \text{cm} \quad \text{liter}}$$

Ans. c = 0.0322 x 10^{-3} M or 0.0322 mM. A 0.0322 mM solution is equalto 0.0322 mmoles per liter or 0.0322 μmoles per ml.

Sometimes the mM extinction coefficient is given. For NADPH this would be 6.22 and would be the absorbance of a 1mM solution of NADPH. If 6.22 is used, then the units of c would be millimolar (μmoles per ml) rather than molar (mmoles per ml). To convert this to μmoles per min of NADPH produced, use the millimolar extinction coefficient of 6.22, i.e., 0.01 = 6.22 x c.

9. The absorbance change was 0.01 per min. The amount of NADPH produced per min (c) is therefore 1.6 x 10^{-3} μmoles per ml. If the volume in the cuvette was 3 ml, then 4.8 x 10^{-3} μmoles of NADPH were produced per minute.

Ans. The number of units of enzyme in the cuvette was 4.8 x 10^{-3}.

10. A = ecl. Therefore, 0.500 = ecl. Since the molecular weight of UTP is 586, the molarity in the solution was 0.029 ÷ 586 = 5 x 10^{-5} M. Therefore, e = 0.500 / 5 x 10^{-5}.

Ans. The molar extinction coefficient is 1 x 10^{-4}.

APPENDIX H

INDEPENDENT PROJECTS

This section consists of independent projects that students may do using some of the techniques learned during the laboratory experiments. Although some guidelines are given as to how to proceed with the projects, the directions differ from those given for the laboratory experiments in being intentionally incomplete in order to give the students experience in designing experiments. It is expected that the students will examine the references cited for the projects and design their own protocols based upon their reading and consultation with the instructor. For example, the student may have to do preliminary experiments to determine how much substrate or inhibitor to add or how long to run an assay. These projects therefore resemble research projects rather than laboratory exercises. Each project should occupy from one to three laboratory periods, depending upon the experiment.

Determing the amount of inorganic phosphate that limits the growth yields in *E. coli*

When *E. coli* is grown under conditions of inorganic phosphate limitation, it responds by making increased amounts of alkaline phosphatase which is a periplasmic enzyme. This was studied in Experiment 11. In order to assay for alkaline phosphatase it is necessary to grow *E. coli* until exogenous inorganic phosphate has been depleted so that the alkaline phosphatase is induced.

To determine the amount of inorganic phosphate that limits growth, grow the cells in a minimal salts medium containing glucose, and add varying amounts of inorganic phosphate. For example, you can use SMB*, (without phosphate buffer), 0.4% glucose, and 10 mM morpholinepropanesulfonic acid (MOPS) buffer, pH 7.4. Make a 0.05 M solution of K_2HPO_4 and add a range of concentrations e.g., 0.05, 0.1, 0.2, and 0.5 mM. Autoclave everything together except the glucose. Dilute a sterile 20% glucose solution to 0.4% into the medium after sterilization. Grow the cells shaking in a Klett flask at 37°C and take absorbancy measurements. Inoculate the flasks from a plate culture or use an overnight culture grown in regular SMB* containing phosphate buffer, at a 1000 fold dilution. You can grow the cells to a low turbidity, refrigerate them, and continue the growth curve the next day.

Factors that affect the relative amounts of saturated, unsaturated, and branched chain fatty acids

Temperature
E. coli may be grown in mineral medium over a temperature range of from 11 to 42°C. The ratio of saturated to unsaturated fatty acids (S/US) in cells (membranes) varies considerably over this range. The effects of temperature were studied by Marr and Ingraham.[1]

Alcohol
Alcohols or other water-miscible solvents that reduce the polarity of water affect the hydrophobic interactions that stabilize membranes and membrane-resident proteins. Addition of alcohols to the medium for growth of bacteria

results in perturbation in the ratio of the saturated/unsaturated fatty acids (S/US) in a way superficially similar to temperature. Read the article by Ingram if you would like to test the effects of alcohols.[2]

Stage of growth

As cells enter the stationary phase the unsaturated fatty acids in cell membrane lipids are converted to the corresponding cyclopropane acids by methyl addition from S-adenosyl methionine. A timed series of cultures analyzed for fatty acids over a 5-6 h period following attainment of stationary phase should show this phenomenon.

Hydrogenation and ozonolysis[3]

Hydrogenation (PtO catalyst) coverts the unsaturated fatty acids to their saturated counterparts. The position of double bonds (16:1 and 18:1) may be determined by oxidation of the sample (e.g., ozonolysis) and gas chromatographic identification of the mono- and dicarboxylic acids that result following methylation.

ENDNOTES

1. Marr, A. G, and Ingraham, J. L. 1962. Effect of temperature on the composition of fatty acids in *Escherichia coli*. *J. Bacteriol.* **84**:1260-1267.

2. Ingram, L. O. 1976. Adaptation of membrane lipids to alcohols. *J. Bacteriol.* **125**:670-678.

3. Ackman, R. G., Sebedio, J.-L., and Ratnayake, W. N. 1981. Structure determinations of unsaturated fatty acids by oxidative fission. *Methods in Enzymology.* **72**:253-276.

Measuring the K_M and V_{max} of fructose-1,6-bisphosphatase and the inhibition by AMP

Refer to Expt. 13 for the preparation of fructose-1,6-bisphosphatase from yeast and to Expt. 12 for a description of how the K_M and V_{maax} can be measured for threonine deaminase. Design an experiment where the fructose-1,6-bisphosphate concentrations are varied from 0 to 2 mM. Remember that the K_M should be in the micromolar range. Plot the data as v vs S and $1/v$ versus $1/S$.

Examine the inhibition by AMP. Use a fructose-1,6-bisphosphate concentration of 0.1 mM and measure the enzyme velocity with different concentrations of AMP from 0 to 0.5 mM. Be certain to test the AMP at concentrations much lower than the substrate concentrations (e.g., between 0 and 100 μM). Determine whether the inhibition is competetive with respect to substrate.

Assaying glycolytic enzymes and fructose-1,6-bisphosphatase in yeast grown on glucose or ethanol

In this project the student determines whether a glycolytic or gluconeogenic carbon source for growth makes any difference with respect to the levels of glycolytic enzymes or fructose-1,6-bisphosphatase in yeast cell extracts. Similar experiments were reported by Foy and Bhattacharjee.[1] Refer to Expt. 13 for conditions of growth and preparation of cell-free extracts. *S. cerevisiae* is grown on yeast extract and peptone containing either 1% glucose or 1% ethanol. The reactions may be assayed at room temperature.

Assay for fructose-1,6-bisphosphatase
Refer to Expt. 13.

Assay for phosphofructokinase

Make a reaction mixture containing the following: 1 mM fructose-6-phosphate (free of glucose-6-phosphate), 5 mM MgCl$_2$, 25 mM potassium phosphate buffer, pH 6.5, 0.15 mM NADH, 0.2 units of aldolase (fructose-1,6-bisphosphate D-glyceraldehyde-3-phosphate lyase, EC4.1.2.13), 1 unit of α-glycerophosphate dehydrogenase (L-glycerol-3-phosphate:NAD oxidoreductase, EC 1.1.1.8), 3 units of triosephosphate isomerase (D-glyceraldehyde-3-phosphate-ketol-isomerase, EC 5.3.1.1), 0.25 mg of supernate protein in a total volume of 1.8 ml. Start the reaction by adding 0.2 ml of 10 mM GTP. (You may have to double the amounts for the colorimeter or spectrophotometer that you are using.) The control does not receive GTP. Follow the reaction by monitoring the decrease in A_{340} as the NADH is oxidized by the dihydroxyacetone phosphate produced by the aldolase that uses the fructose-1,6-bisphosphate as its substrate.

Assay for phosphoglucose isomerase

Prepare a reaction mixture containing: 0.05 mM NADP$^+$, 0.3 units of glucose-6-phosphate dehydrogenase(D-glucose-6-phosphate:NADP1-oxido-reductase, EC 1.1.1.49), 50 mM triethanolamine-HCl buffer containing 10 mM MgCl$_2$ adjusted to pH 7.4 with 2 N NaOH, and 0.25 mg of supernate protein to a final volume of 1.8 ml. Start the reaction by adding 0.2 ml of 10 mM fructose-6-phosphate. Monitor the reaction by following the increase in A_{340} as the NADP$^+$ becomes reduced by the glucose-6-phosphate generated in the reaction.

Regulation of phosphofructokinase activity

Phosphofructokinase is an allosteric enzyme that catalyzes a reaction that is a key control point in glycolysis and has often been referred to as a pacemaker enzyme for that pathway. Allosteric effectors include ATP, AMP, and citrate. In this project you will investigate the regulation of phosphofructokinase from yeast. Refer to Expt. 13 for growing the yeast and making the extracts. The assay conditions for phosphofructokinase are described earlier in this section. Use ATP rather than GTP.

Assay the enzyme at different substrate concentrations and plot v versus S. You will have to use a concentration of ATP that is not inhibitory at the substrate concentrations used. You should obtain a Michealis-Menten type curve. Do the same assay but in the presence of an inhibitory concentration of ATP. Determine the K_M and V_{max} with and without ATP.

It has been reported that the inhibition by ATP is reversed by AMP, and that citrate enhances the inhibition by ATP. Determine whether this is the case for the enzyme that you are assaying.

Further experiments and demonstrations with *Photobacterium*

1. A cell suspension of *Photobacterium* prepared as for Expt. 16 can be used to show a class the dependence of light production on substrate, O_2, and the effect of an uncoupler or cyanide (but not both of these) by placing the suspension (warmed to room temperature) in a large test tube or cylinder in dim light and adding substrate (glycerophosphate or complete broth medium), agitating (with intermittent periods without shaking) to show the effect of O_2, and, finally, by addition of the inhibitor/uncoupler.

2. Ethical considerations that make the testing of drugs, cosmetics, etc. for safety by use of animals have impelled a search for microbial surrogate systems (e.g., the Ames test for mutagenic activity) for use in such testing. One such system employs *Photobacterium* and its light production. Devise ways to test various substances for toxicity using this system. Can you think of how to make the test quantitative?

3. The strict dependence of light production by *Photobacterium* on oxygen enables the use of cell suspensions of this bacterium as a sensitive indicator of O_2 production. Mixing a glowing suspension that has been allowed to go anaerobic with a marine alga (single celled or a multicellular seaweed) for instance, in a glass vessel and exposing the vessel to an intense pulse of light (from a photoflash lamp) provides a brief flash of bacterial light for several seconds following the flash.

4. The development of the luminous system of *Photobacterium*, like an increasing number of microbial traits, has been shown to be dependent on the cell density sensed via an excreted pheromone molecule (typically a homoserine lactone). This dependence can be demonstrated by measuring light production by a culture during growth. Light production typically begins only in the late log phase of growth. Use *Photobacterium* broth (with $CaCO_3$ omitted for such liquid cultures) and plot intensity of light production and culture turbidity vs. time on the same graph.